Dale M. Johnson

Professor of Research
The University of Tulsa
Tulsa, Oklahoma

ISBN: 0-538-60016-0

COVER PHOTO: Cover Photo by Melvin L. Prueitt
Los Alamos National Laboratory

1 2 3 4 5 6 7 8 PR 5 4 3 2 1 0 9 8

Printed in the United States of America

Preface

The purpose of this study guide is to assist students in refining their knowledge and understanding of statistical content presented in the textbook. It is intended to be used as a supplement to the text:

PROBABILITY AND STATISTICS by Dale M. Johnson
(Cincinnati: South-Western Publishing Co., 1989).

The structured format provides an opportunity for students to diagnose their strengths and weaknesses, practice computational and analytical skills, and reinforce statistical concepts and terminology presented in the textbook.

The chapters in this study guide parallel the content of the respective chapters in the main text. The content in each study guide chapter is divided into four sections:

1. Concept Comprehension
2. Vocabulary Practice
3. Rehearsal Exercises
4. Practice Problems

The Concept Comprehension sections are narrative reviews of the chapters and call for students to fill in blanks with important concepts (words, phrases, or symbolic notation). Each chapter provides practice with statistical vocabulary in the Vocabulary Practice section. This is accomplished in a crossword-puzzle format. The Rehearsal Exercises sections provide opportunities to apply the concepts in the chapter and consist of multiple-choice questions. Each chapter is terminated with a Practice Problems section. Aids are sometimes provided to help structure solutions. Items from optional material in the textbook are marked with an asterisk (*).

Ongoing use of the study guide along with the text and class activities is encouraged. It will assist students in their attempt to master the current content and should also prove useful for review purposes at the conclusion of each chapter or prior to examinations. Therefore the study guide is viewed as a flexible complementary learning aid and review source for the text.

Dale M. Johnson

Contents

CHAPTER 1
Introduction to Statistics

Student ______________________________ Total Score __________

SECTION 1 CONCEPT COMPREHENSION

Directions: Complete the following by writing the correct word or words in the respective blanks to the right.

Answers **For Scoring**

___(1)___ is the science dealing with the organization, summary, analysis, and interpretation of numerical data. In its plural and more common usage form, it is often used to mean ___(2)___. The data originate from the assignment of numbers, according to well-defined rules, to objects, attributes, performances, and characteristics, which is the process of ___(3)___. Generally the first task facing the statistician is to ___(4)___ the data into tables, graphs, or other orderly arrangements. The data then may be ___(5)___ to provide an abbreviated description of the entire set of numbers. The data are analyzed, and the final step is to ___(6)___ the results.

Data may be used simply to identify or name individuals or groups of individuals when used on a ___(7)___ scaling level. Other classifications of scaling levels are ___(8)___, ___(9)___, and ___(10)___. Only ___(11)___ scaling permits one to measure quantities in absolute amounts.

A measurable characteristic for which values differ is called a(n) ___(12)___. Variables that generally take on integer values are called ___(13)___, whereas variables that theoretically can assume any value within the measurable limits are referred to as being ___(14)___.

Statistical predictions must be interpreted under conditions and rules governed by ___(15)___. Thus statistical results do not unconditionally ___(16)___ any point with certainty.

Tabulating, organizing, graphing, and summarizing numerical data are generally thought of as methods included in one of the two broad classifications of statistics known as ___(17)___ statistics. The other broad classification, which deals with data that are only partially available (that is, sample data) to researchers for the purpose of estimating and hypothesis testing, is called ___(18)___ statistics.

1. __________ 1. ____
2. __________ 2. ____
3. __________ 3. ____
4. __________ 4. ____
5. __________ 5. ____
6. __________ 6. ____
7. __________ 7. ____
8. __________ 8. ____
9. __________ 9. ____
10. __________ 10. ____
11. __________ 11. ____
12. __________ 12. ____
13. __________ 13. ____
14. __________ 14. ____
15. probability 15. ____
16. prove 16. ____
17. __________ 17. ____
18. __________ 18. ____

SECTION 2 VOCABULARY PRACTICE

Directions: Identify the term defined and complete the designated squares.

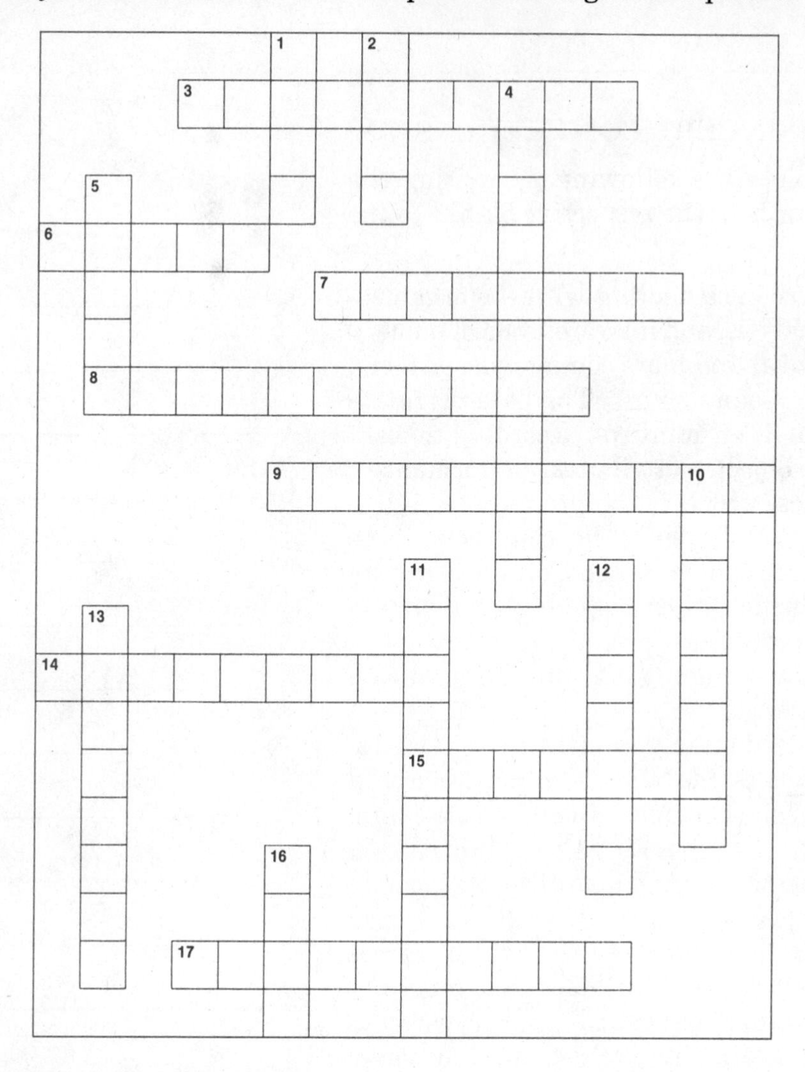

Across **For Scoring**

3. The science dealing with organizing, summarizing, analyzing, and interpreting numerical data. ________
6. The level of scaling that reveals the absolute amount of a trait, characteristic, or performance. ________
7. Data that may assume only whole number values. ________
8. The process of assigning a number to a trait or performance according to a specified rule. ________
9. The branch of statistics that deals with organizing and summarizing data that have been collected. ________
14. Within the context of the definition of statistics as a science, calculations using statistical formulas would be ______ data. ________
15. The numerical scaling that permits only naming or labeling individuals or groups. ________
17. A variable that theoretically can assume any real value within the range. ________

(continued)

Student ______________________________ Score __________

SECTION 2 VOCABULARY PRACTICE (continued)

Down **For Scoring**

1. Statistics is the study of numerical ______. ________
2. The term popularized in computer science that means that the quality of results is dependent on the quality of the information used to obtain the results. ________
4. Descriptive and ______ are two major branches of statistics. ________
5. The singular form of data. ________
10. A measurable characteristic that assumes different values. ________
11. Constructing a table or graph is a method of ______ data. ________
12. The level of scaling referred to as "rank ordering." ________
13. The level of scaling that permits differences between individuals to be determined but does not reveal how much of an attribute is present in an absolute sense. ________
16. If a person finishes third in a track-meet event, 3 represents her or his ______ in the event. ________

SECTION 3 REHEARSAL EXERCISES

Directions: In the Answers column, write the letter that represents the correct choice.

Answers **For Scoring**

1. Which of the following is not considered a function of statistics? ______ 1. ______
   (A) organizing data
   (B) describing data
   (C) unconditionally proving a conclusion
   (D) estimating values
2. Assigning numbers to objects, attributes, or performances according to specified rules best defines: ______ 2. ______
   (A) hypothesis testing
   (B) inferential statistics
   (C) descriptive statistics
   (D) measurement
3. Which of the following would best illustrate a discrete variable? ______ 3. ______
   (A) the weights of the backfield on the football team
   (B) the number of brothers and sisters of senior students
   (C) the amount of time required for students to pass from one class to another
   (D) the area of classrooms in your school in square feet
4. An "average" of a set of numbers illustrates the concept of: ______ 4. ______
   (A) summarizing data
   (B) organizing data
   (C) nominal scaling
   (D) discontinuous data

(continued)

SECTION 3 REHEARSAL EXERCISES (continued)

	Answers	For Scoring
5. Statistical results and reports are consumed: (A) almost exclusively by college/university professors (B) primarily by politicians (C) by virtually everyone (D) by nearly no one	______	5. _____

SECTION 4 PRACTICE PROBLEMS

Use the following notation to mark the level of measurement scaling implied by statements 1-10:

N = Nominal, O = Ordinal, I = Interval, and R = Ratio.

	Answers	For Scoring
1. In the regional debate tournament, Max was first, Donald was second, and Philip was third.	______	1. _____
2. Each student in school was assigned a student identification (ID) number.	______	2. _____
3. Shirley is 2 inches taller than Marian.	______	3. _____
4. Sandra finished her exam in 32 minutes while Julie finished in 25 minutes.	R	4. _____
5. The split end on the football team wears jersey number 87.	______	5. _____
6. Sam ranked fifth in the senior class in achievement.	______	6. _____
7. The temperature in the chemistry lab was 92° Fahrenheit.	I	7. _____
8. The weight of 1 cc of water is about 1 gram.	______	8. _____
9. Julio is $2\frac{1}{2}$ times as old as his younger sister, who is three years old.	______	9. _____
10. The address of Jackson County High School is Route 5, Box 313.	______	10. _____

11. In a certain census area, it was found that 204 of 2,040 households had incomes between $30,000 and $80,000 per year.

	Answers	For Scoring
a. What percentage of the households had annual incomes in the range $30,000 to $80,000?	10%	11a. _____
b. What is your guess as to what percentage of the households had annual incomes between $30,000 and $40,000?	2%	11b. _____
c. What do you guess would be the percentage of households with incomes between $30,000 and $50,000 per year?	4%	11c. _____
d. What assumption are you making about the distribution of incomes within the range $30,000 to $50,000?		11d. _____

____

____

____

(continued)

Student ______________________ Score ______

SECTION 4 PRACTICE PROBLEMS (continued) **For Scoring**

12. Mrs. Matthews, a new algebra teacher at McGuire High School, found that her students scored an average of 84% correct on exams throughout the year and that only 2% failed the course. She concluded that these data were proof that she was an exceptionally effective teacher. Is she justified in her claim? Comment. 12. ______

13. During the mid-1970's, the first inoculations in a nationwide program to test the effectiveness of swine flu vaccine were given to high-risk groups, including the elderly. Of more than 20,000 persons over age 65 who took the vaccine, 3 died. Because of the three deaths, several states suspended the vaccination programs. From the standpoint of a researcher, comment on the justification of the suspension by the states. 13. ______

14. The following statements are some "statistical" comments taken from the opening section of the text. For each case, comment on your interpretation of the meaning.

    a. ". . . they averaged 5.2 yards per play . . ." 14a. ______

    b. ". . . the crowd was larger than normal . . ." 14b. ______

    c. ". . . scoring at the 98th percentile . . ." 14c. ______

    d. ". . . the mean of the control group was 8.72 . . ." 14d. ______

(continued)

For Scoring

15. In one city-wide poll, 56% of those responding gave Mayor Allen high marks; the percentage in another poll was 52%. Speculate on what could have caused the 4-percentage-point difference.

15. ____

16. Several years ago a newspaper advice columnist was asked whether having children was worth the trouble involved. The columnist in turn asked her readers to comment. Several weeks later the headline on her column was "70% of Parents Say Kids Not Worth It" because 70% of those readers who were parents and who responded indicated they would not have children if they could make the choice again. Comment on this experience as a *valid* inference about the general population of parents.

16. ____

17. Suppose that a certain group of high-school seniors are selected to be interviewed to provide information about their feelings regarding the effectiveness of the teaching at their school. Would the resulting data be the same if the interviews were carried out by teachers; or by school administrators; or by parent volunteers; or by other students? Comment on obvious biases.

17. ____

18. A sales firm operating in Denver uses a device that randomly dials residential telephone numbers in that city. Suppose that 650 of the first 1,000 calls were to residential numbers having family members under 21 living at home. According to the census, 68% of the residential units have family members under 21 living at home. Is this degree of discrepancy surprising? Why?

18. ____

CHAPTER 2

Displaying Data

Student ______________________________ Total Score __________

SECTION 1 CONCEPT COMPREHENSION

Directions: Complete the following by writing the correct word or words in the respective blanks to the right.

Organizing data in a visual display so that the lengths of rectangles represent quantities produces a chart called a(n) __(1)__ graph. When constructing such a graph, one explicitly shows the discrete nature of one of the variables by placing __(2)__ between the bars. __(3)__ bar graphs show the lengths of the bars across the graph, whereas in __(4)__ bar graphs, bars are placed up and down relative to the viewer. When two or more subcategories of a particular level of a discrete variable are shown side by side, the extension of a simple bar graph is called a(n) __(5)__ bar graph. Comparisons of the cumulative values of such subcategories can be shown in a __(6)__ bar graph.

The circle or __(7)__ graph represents various fractional parts of the whole. The sizes of the central __(8)__ are proportional to the respective quantities represented. Thus if one-quarter of the whole is represented, an angle of __(9)__ degrees would be constructed to represent the appropriate fractional part of the circle graph.

While most forms of graphs portray reasonably accurate accounts when studied carefully, first impressions may lead to __(10)__ conclusions about the data because of purposeful or unintentional creations of illusions. For example, the proportional lengths of bars on a bar graph may be distorted by failure to start the scaling of all bars at the __(11)__ point. __(12)__ may exaggerate differences because of the implied three-dimensional illusion created by the images used to represent various quantities. These generally attract attention and can be misleading at first glance. Quantities should be represented by the *number* of images or icons, not by using varying __(13)__ of images.

	Answers	For Scoring
1.	______________	1. ____
2.	______________	2. ____
3.	______________	3. ____
4.	______________	4. ____
5.	______________	5. ____
6.	______________	6. ____
7.	______________	7. ____
8.	______________	8. ____
9.	______________	9. ____
10.	______________	10. ____
11.	______________	11. ____
12.	______________	12. ____
13.	______________	13. ____

SECTION 2 VOCABULARY PRACTICE

Directions: Identify the term defined and complete the designated squares.

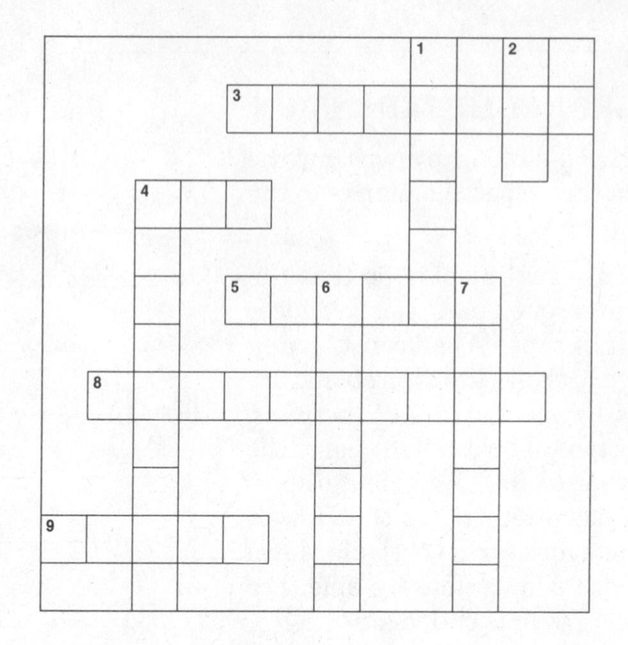

Across

For Scoring

3. A ______ bar graph has the bars running up and down the page relative to the observer. ______
4. A circle graph. ______
5. The ______ in a circle graph are proportional in size to the quantities they represent. ______
8. On a vertical bar graph, the levels or categories of the nominal (or discrete) variable are shown on the ______ axis. ______
9. A visual representation of organized data. ______

Down

1. A ______ graph divides the various categories into segments that total 360 angular degrees. ______
2. ______ graphs use vertical or horizontal rectangles to represent quantities. ______
4. A graph that uses icons or figures to represent quantities. ______
6. A bar graph in which the subcategories are compared side by side. ______
7. A bar graph in which the subcategories are accumulated to form the lengths of the bars. ______

Student ______________________________ Score __________

SECTION 3 REHEARSAL EXERCISES

Directions: In the Answers column, write the letter that represents the correct choice.

For exercises 1 through 5, refer to the graph that follows, which shows the number of correct responses by boys and girls on a 10-problem exam.

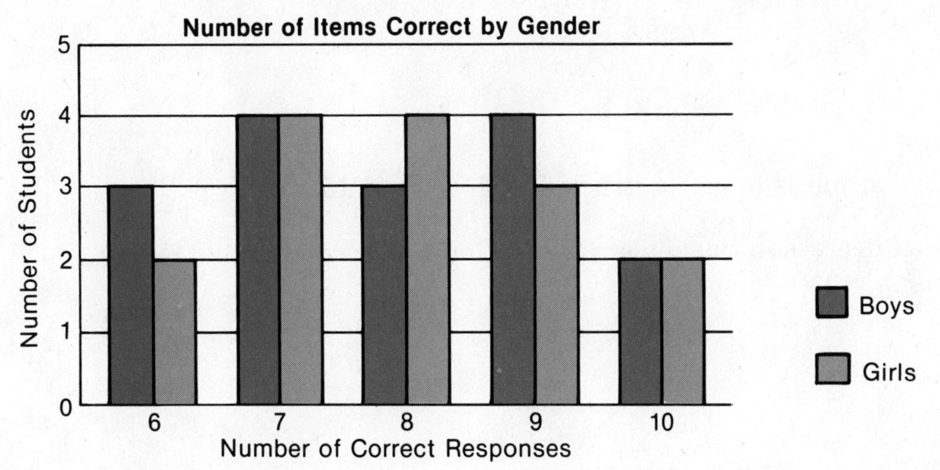

	Answers	For Scoring
1. This graph is called a: (A) horizontal bar graph (B) stacked bar graph (C) grouped bar graph (D) pictogram	______	1. _____
2. How many boys correctly answered six problems? (A) 1 (B) 2 (C) 3 (D) none of the above	______	2. _____
3. Assuming that all class members correctly answered six or more of the items, how many girls are in the class? (A) 4 (B) 10 (C) 12 (D) 15	______	3. _____
4. How many items were correctly answered by the largest number of students (boys and girls)? (A) 7 (B) 8 (C) 9 (D) 11	______	4. _____

(continued)

	Answers	For Scoring
5. How many boys correctly answered nine or more items?	______	5. ______

(A) 4
(B) 6
(C) 9
(D) 11

For exercises 6 through 10, use the data that follow, which represent the number of students participating in a charity walk by grade level.

Grade Level	Number of Participants
Freshman	15
Sophomore	30
Junior	45
Senior	60

Assume that a pie graph is to be constructed with these data.

6. The number of degrees in the angle representing seniors on the pie graph equals: ______ 6. ______
(A) 60
(B) 144
(C) 166.67
(D) 360

7. The proportion (as a two-decimal-place fraction) of freshmen in the group equals: ______ 7. ______
(A) 0.04
(B) 0.10
(C) 0.15
(D) 0.36

8. How does the angle representing the seniors compare to the angle representing the sophomores? ______ 8. ______
(A) The angle for the seniors is 30 degrees larger.
(B) The angle for the seniors is 144 degrees larger.
(C) The angle for the seniors is twice as large.
(D) The angle for the seniors is three times as large.

9. The sum of the angles representing all four grade levels equals: ______ 9. ______
(A) 90 degrees
(B) 100 degrees
(C) 150 degrees
(D) 360 degrees

10. Approximately how many angular degrees are represented by each individual student? ______ 10. ______
(A) 2.4
(B) 1.0
(C) 0.15
(D) 0.42

Student ______________________ Score __________

SECTION 4 PRACTICE PROBLEMS

For Scoring

1. Given the following data on computer use in high schools by subject area, construct a vertical bar graph on the axes provided. 1. _____

Subject	Percentage of Class Using Computers
Business	72
Mathematics	85
Physical Science	70
English and Language Arts	65
Social Science	54
Other	40

%
90
80
70
60
50
40
30
20
10
0

Business | Mathe-matics | Physical Science | English and Language Arts | Social Science | Other

(continued)

For Scoring

2. Sources and amounts of income for a school booster club were as follows:

Club-sponsored car wash	$390
Profits from concession at sporting events	$1,160
Community donations	$640
Other	$690

What central angles in a pie graph would represent the four incomes?

a. car wash $\frac{390}{2,880} \times 360$ degrees = __________ 2a. ______

b. concession = __________ 2b. ______

c. donations = __________ 2c. ______

d. other = __________ 2d. ______

3. Students were surveyed and asked which of five soft drinks they preferred to be available in the school vending machine. The results (in percentages, by gender) were as follows: 3. ______

Soft Drink	Male (%)	Female (%)
Brand A	20	15
Brand B	25	15
Brand C	15	30
Brand D	18	25
Brand E	22	15

Construct a vertical grouped bar graph displaying the data.

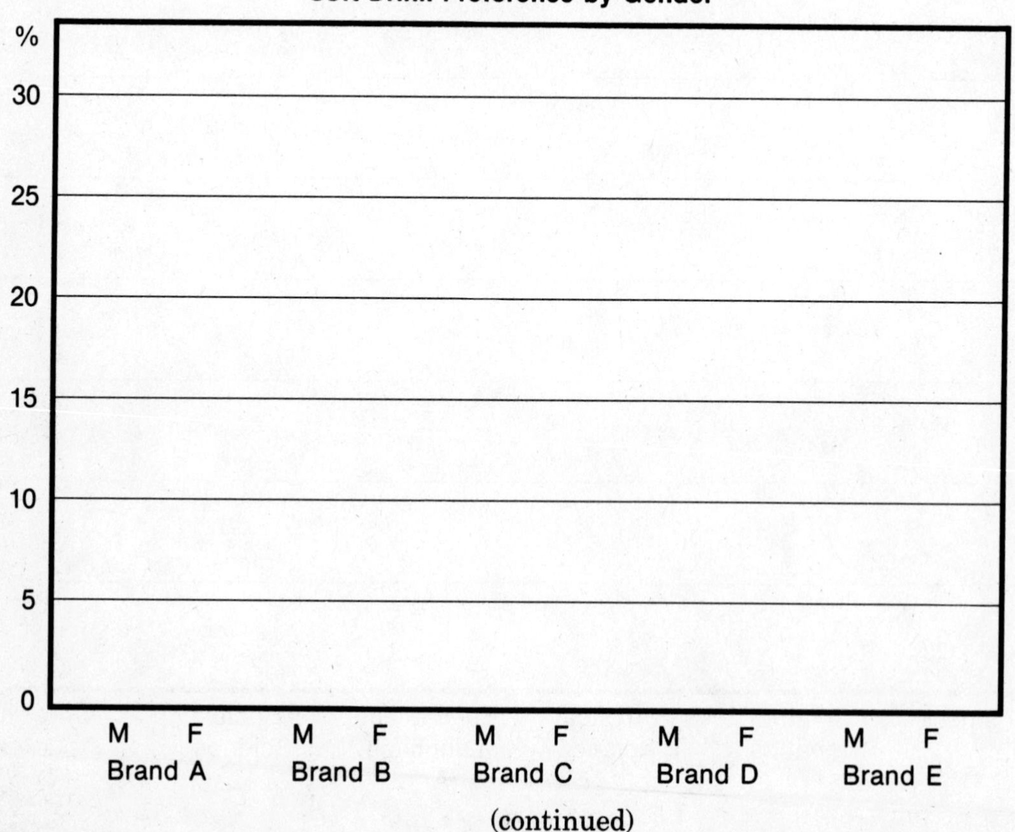

(continued)

Student ______________________________ Score ________

SECTION 4 PRACTICE PROBLEMS (continued)

For Scoring

4. Using the data in problem 3, construct a vertical stacked bar graph. 4. ____

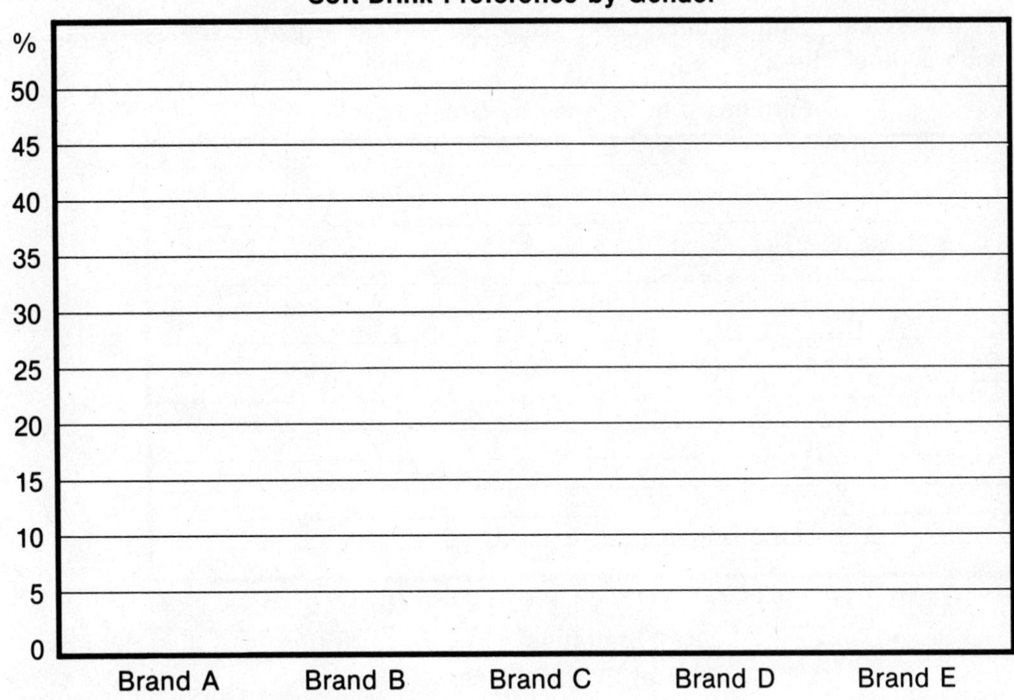

5. Using the data from problem 3, sketch a horizontal pictogram of the male percentages using an appropriate image representation. 5. ____

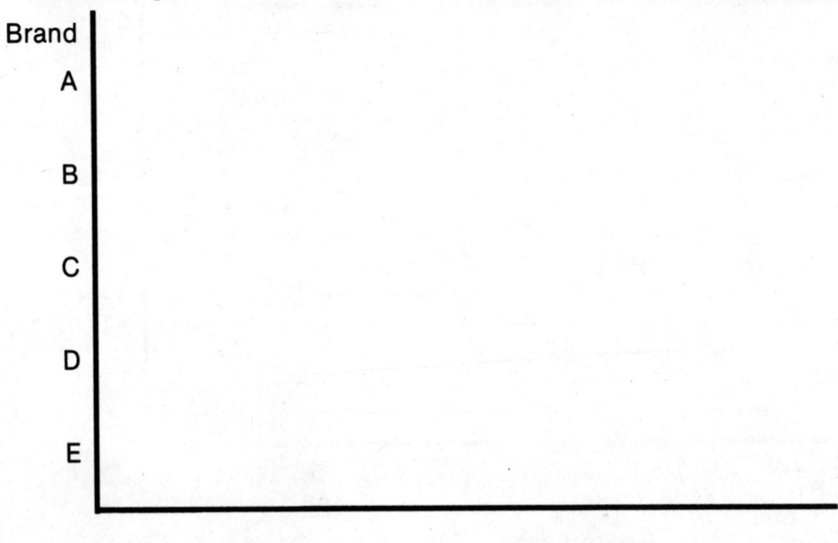

Each represents %

(continued)

For Scoring

6. The following data show a metropolitan high school's enrollment in aerobic dance, statistics, and creative writing classes by grade level.

Grade Level	Number Enrolled in Aerobic Dance	Statistics	Creative Writing
10	20	12	20
11	16	18	19
12	18	24	15

a. Construct a vertical grouped bar graph grouping the three grade levels within each type of class. 6a. ______

Enrollment in Classes by Grade Level

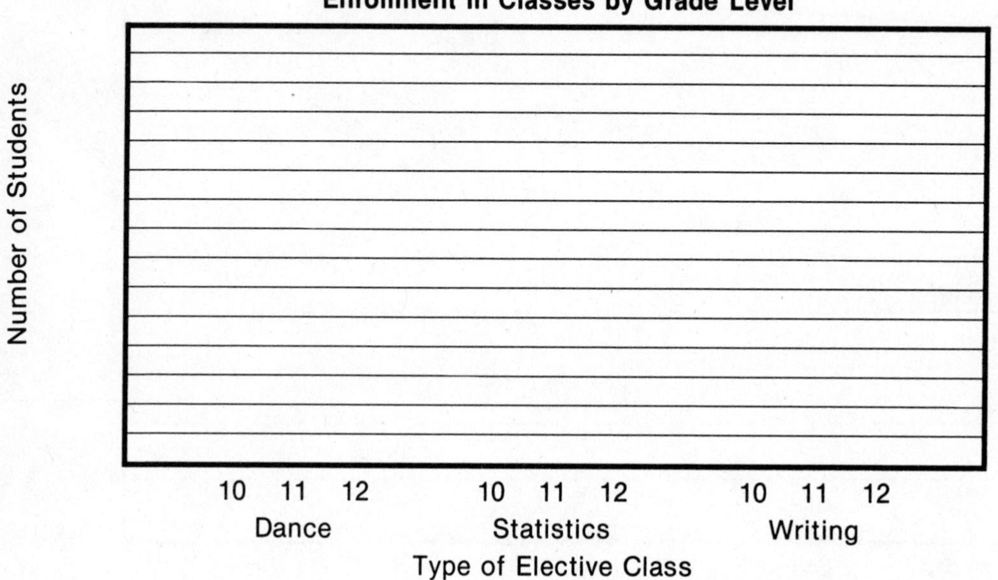

b. Construct a vertical grouped bar graph grouping the three types of classes within each grade level. 6b. ______

Enrollment by Type of Class

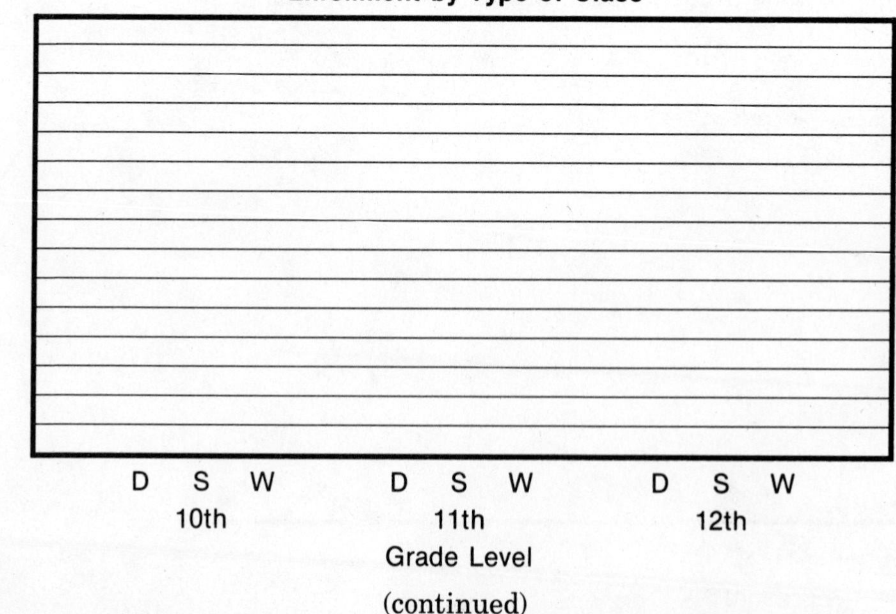

(continued)

Student ______________________________ Score __________

SECTION 4 PRACTICE PROBLEMS (continued)

For Scoring

c. Describe the differences in the two graphs a. and b. That is, although they both display identical data, how do their respective functions differ?

6c. _____

____

____

____

____

____

7. Using the data from problem 6, construct three circle graphs showing the enrollment by grade level in each of the three types of classes.

a. Aerobic Dance

7a. _____

Enrollment in Aerobic Dance by Grade Level

Grade	Central Angle °
10th	________
11th	________
12th	________

b. Statistics

7b. _____

Enrollment in Statistics by Grade Level

Grade	Central Angle °
10th	________
11th	________
12th	________

(continued)

For Scoring

c. Creative Writing

7c. ______

Enrollment in Creative Writing by Grade Level

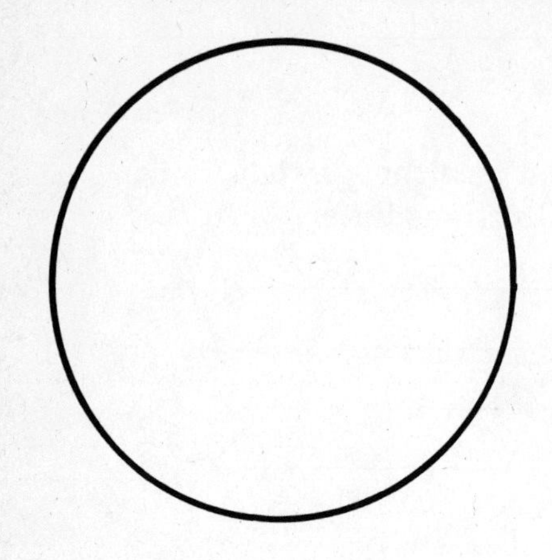

Grade	Central Angle °
10th	______
11th	______
12th	______

8. Draw a vertical grouped bar graph for the following data, which show the number of winners and second-place finishers in the high-school women's 100-yard dash in the respective lane assignments during a track season.

8. ______

Lane No.	No. of Winners	No. of Second Places
1	14	10
2	10	8
3	12	10
4	18	17
5	14	16
6	9	14
7	11	14
8	15	14

Lane Assignments for Women's Track

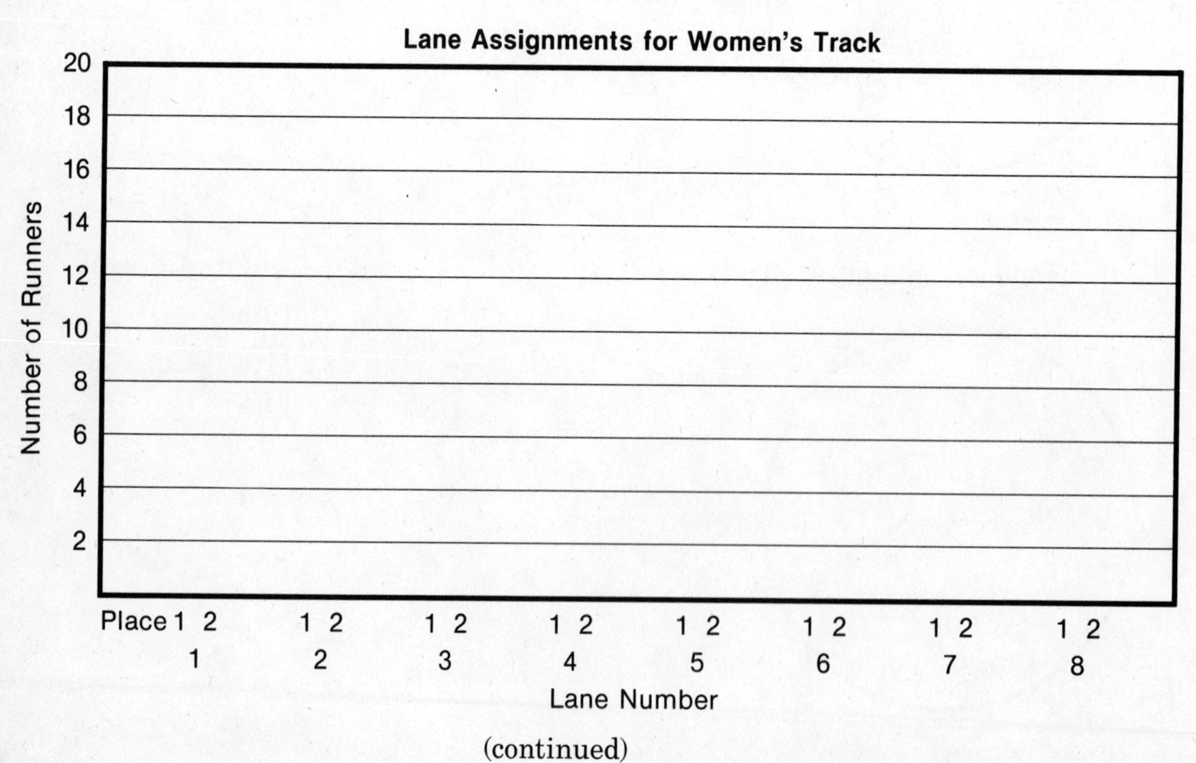

(continued)

Student ______________________ Score ________

SECTION 4 PRACTICE PROBLEMS (continued)

For Scoring

9. Refer to the women's track data in problem 8. Add the number of winners and second-place finishers. Then show the collapsed data on a simple vertical bar graph. 9. ____

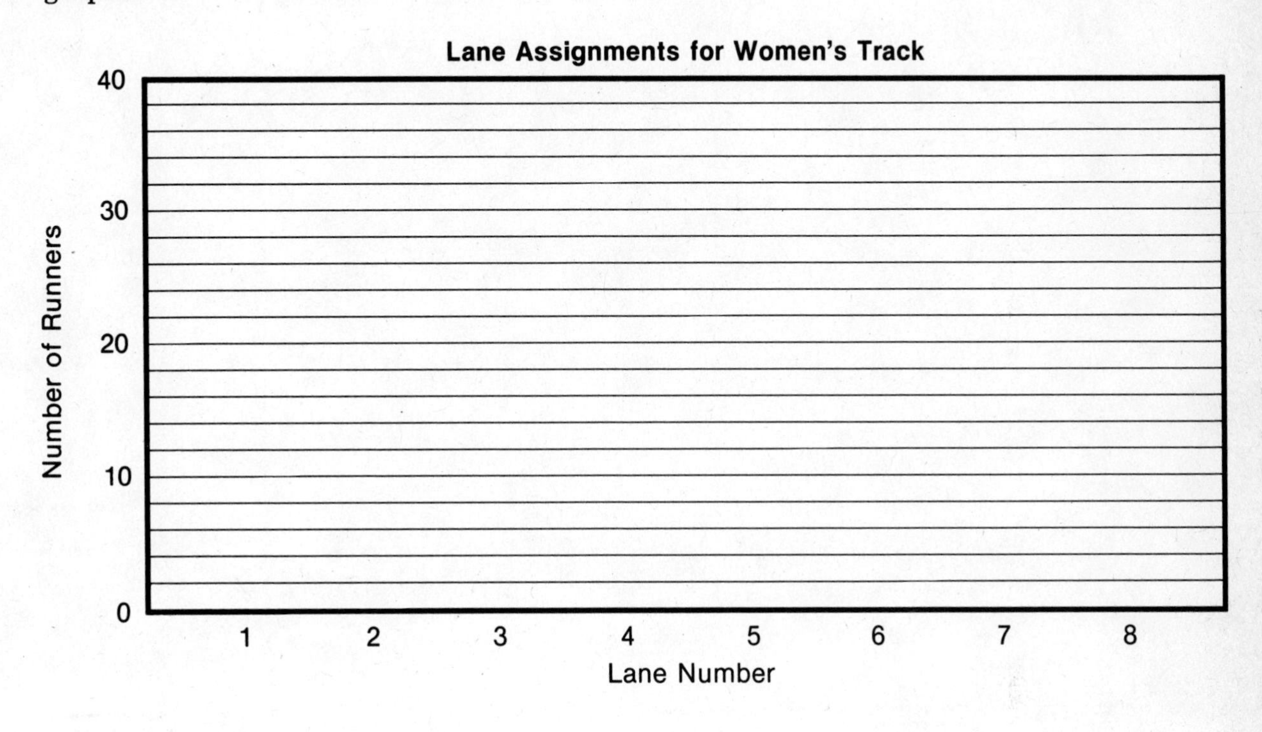

10. Construct a circle graph of the collapsed data that are displayed in problem 9. 10. ____

Frequency Distributions

Student ______________________ Total Score ____________

SECTION 1 CONCEPT COMPREHENSION

Directions: Complete the following by writing the correct word or words in the respective blanks to the right.

When one is first attempting to arrange unordered data in a table so that the number of occurrences of each value in the distribution can be easily viewed, a(n) __(1)__ distribution is appropriate. Individual values of the variables may be used, or the values may be grouped into intervals called __(2)__ intervals. Such a distribution can be readily changed into a relative frequency distribution, which shows the __(3)__ of occurrence of the values. When the relative frequency distribution is accumulated from the low through the high value of the variable, the resulting tabular column is a(n) __(4)__ frequency distribution, and the largest value in this column must be __(5)__.

__(6)__ are bar graphs without spaces (to explicitly show the __(7)__ nature of the variables) showing information identical to the frequency distribution. The __(8)__ and smooth-line curve are alternate forms of the histogram. When a relative cumulative frequency distribution is shown as a smooth-line graph, it is called a(n) __(9)__.

__(10)__ frequency distributions involve two variables. In the case of a two-variable or __(11)__ distribution, a contingency table shows joint frequencies in each __(12)__; and when the rows and columns are summed, the resulting values are known as __(13)__. If the two variables are continuous, a bivariate plot or __(14)__ may be used to visually show the variables simultaneously on the graph.

The __*(15)__ diagram actually uses values of a variable to form a type of horizontal bar graph. If values of a variable are within the range of 0 to 99, the "units" portion of the numbers would serve as __*(16)__, whereas the "tens" place values would serve as __*(17)__.

	Answers	For Scoring
1.	______________	1. ____
2.	______________	2. ____
3.	______________	3. ____
4.	______________	4. ____
5.	______________	5. ____
6.	______________	6. ____
7.	______________	7. ____
8.	______________	8. ____
9.	______________	9. ____
10.	______________	10. ____
11.	______________	11. ____
12.	______________	12. ____
13.	______________	13. ____
14.	______________	14. ____
15.	______________	15. ____
16.	______________	16. ____
17.	______________	17. ____

*Any item that has an asterisk before the number indicates it covers optional text material. If such material was not presented in class, you may skip this item. Check with your teacher.

SECTION 2 VOCABULARY PRACTICE

Directions: Identify the term defined and complete the designated squares.

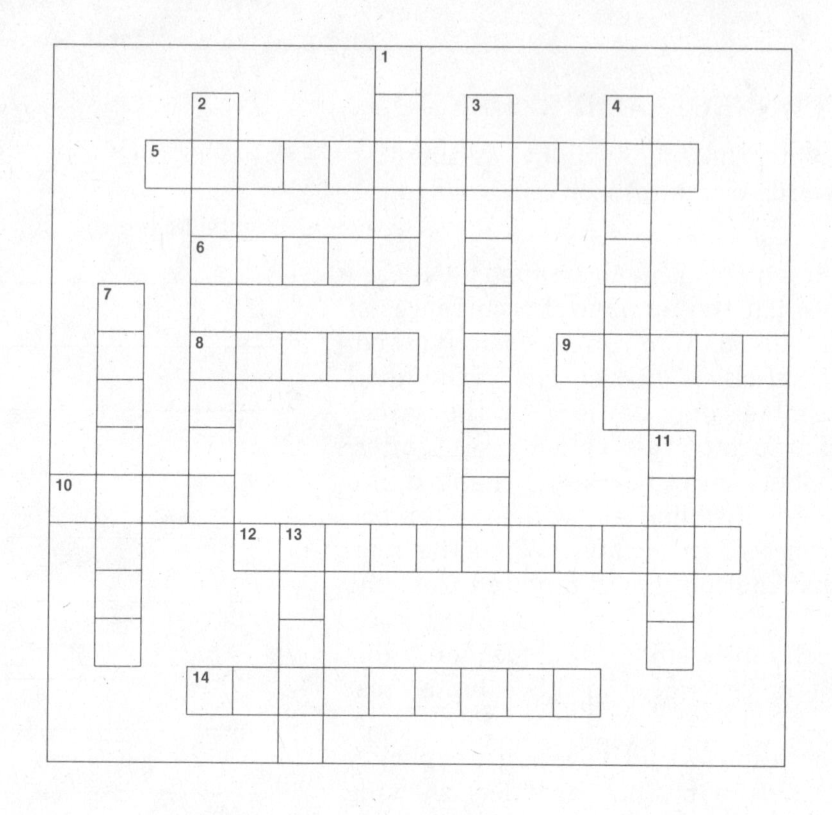

Across

For Scoring

5. An organized set of numerical values is a numerical ______. ______
6. A frequency distribution is organized in a ______ form. ______
8. A visual portrayal of organized data. ______
9. A ______ frequency distribution shows the frequency of two categorical variables in a table. ______

*10. A ______-and-leaf diagram allows the re-creation of the original data. ______

12. A graph that shows the plots of two variables simultaneously. ______
14. Refers to a distribution of two variables. ______

Down

1. A smooth-line graph of a cumulative frequency distribution. ______
2. A bar graph without spaces between the bars. ______
3. When a running total of the frequencies of the values in a distribution is recorded, the resulting ______ frequency distribution shows *the number* of values equal to or below any particular value. ______
4. A broken-line graph of a frequency distribution. ______
7. A ______ frequency distribution expresses frequencies in proportions or percentages. ______
11. An interval of values within a distribution. ______
13. A smooth-line graph. ______

Frequency Distributions

Student ______________________________ Score __________

SECTION 3 REHEARSAL EXERCISES

Directions: In the Answers column, write the letter that represents the correct choice.

Answers | For Scoring

1. A *grouped* frequency distribution is more likely to be used when: ______ 1. ______
   (A) one needs to gain in accuracy and detail
   (B) the distribution has a large number of values
   (C) the range between the low and high scores is large
   (D) both conditions specified in B and C are true

2. The *real limits* of an interval with apparent limits of 740 – 744 are: ______ 2. ______
   (A) 740.5 – 744.5
   (B) 740.0 – 744.0
   (C) 739.9 – 744.1
   (D) 739.5 – 744.5

3. When setting up class intervals, one reason to select interval widths that are odd numbers is so that: ______ 3. ______
   (A) real limits will not have fractions
   (B) the midpoint of the interval will be a whole number
   (C) the number of intervals covering the range will be even
   (D) a more accurate histogram can be constructed

4. A table with one column made up of a continuous variable and the other showing the frequency of occurrence of each measured value of the variable is called a(n): ______ 4. ______
   (A) histogram
   (B) relative frequency distribution
   (C) frequency distribution
   (D) ojive

5. An original data set can be recreated (reconstructed) from a(n): ______ 5. ______
   (A) relative frequency distribution
   (B) relative cumulative frequency distribution
   (C) grouped frequency distribution
   (D) ungrouped frequency distribution

6. If, for a continuous variable, the apparent limits of a class interval are 95 – 99, the *real limits* are: ______ 6. ______
   (A) 94.5 – 99.5
   (B) 94 – 100
   (C) 95.5 – 98.5
   (D) 95.1 – 98.1

(continued)

SECTION 3 REHEARSAL EXERCISES (continued)

Answers For Scoring

7. The most efficient method of showing the *number* of cases falling at or below a particular value of a continuous variable is with a: ______ 7. _____
   (A) relative frequency distribution
   (B) cumulative frequency distribution
   (C) histogram
   (D) frequency polygon

8. A bivariate distribution refers to: ______ 8. _____
   (A) a single measure on two persons or cases
   (B) two measures on each case
   (C) two displays, one a table and one a graph for the set of data
   (D) variables that may take on only two values

9. In a contingency table, the sums of the frequencies in the rows and columns are called: ______ 9. _____
   (A) frequency distributions
   (B) ojives
   (C) bivariate plots
   (D) marginals

10. Assuming each of the following methods was used to display a set of data, which would provide a different type of information from the other three? ______ 10. _____
    (A) cumulative frequency distribution
    (B) ojive
    (C) histogram
    (D) percentile (centile) equivalent

SECTION 4 PRACTICE PROBLEMS

For problems 1 through 7, use the following test raw scores on a 50-item "vocational maturity" scale taken by a group of sophomore students.

Scores

27	33	29	32	30	29	35	31
32	29	30	34	31	35	32	32
30	33	27	31	31	32	28	29
26	33	28	36	31	30	30	32
31	29	32	30	29	33	32	34
33	28	30	34	31	28	32	27
31	28	29	31	32	31	27	29
31	35	30	31	28	32	25	31
30	32	29	33	26	33	31	30
29	33	31	28	30	30	34	28

(continued)

Student ______________________________ Score ________

SECTION 4 PRACTICE PROBLEMS (continued)

For Scoring

1. Complete the following table in the spaces provided using the data just presented. 1. _____

Score	Tally	f	cf	rf	rcf
36					
35					
34					
33					
32					
31					
30					
29					
28					
27					
26					
25					

2. Using 3 as the interval width and a lower apparent limit of 24 for the lowest class, complete the grouped frequency distribution using the same data. 2. _____

Class Interval	Tally	f
36 – 38		
33 – 35		
30 – 32		
27 – 29		
24 – 26		

What are the real limits for the class interval 33 – 35?

____

(continued)

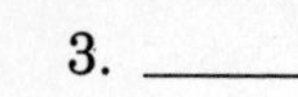

3. Use the results of problem 1 and construct a simple histogram of the frequency distribution. 3. _____

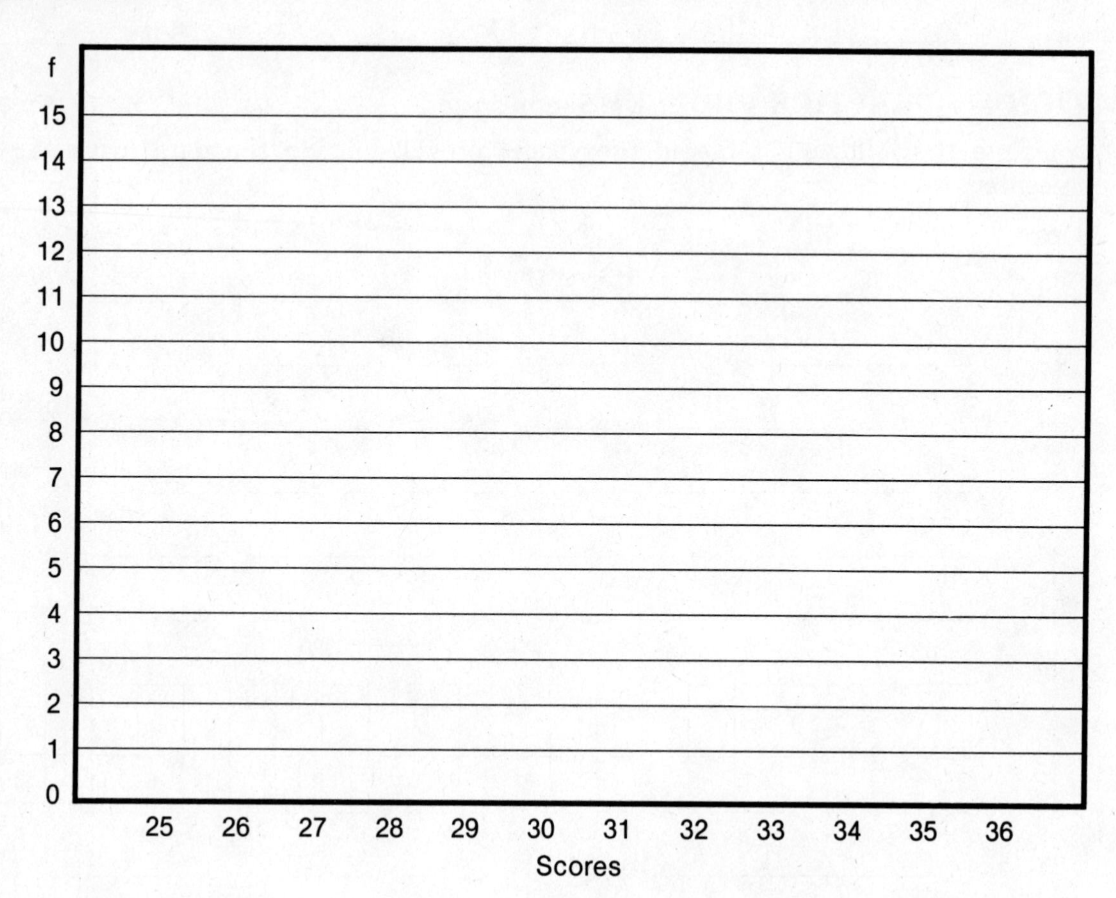

4. By referring to problem 2, construct a histogram using the class intervals along the horizontal axis. 4. _____

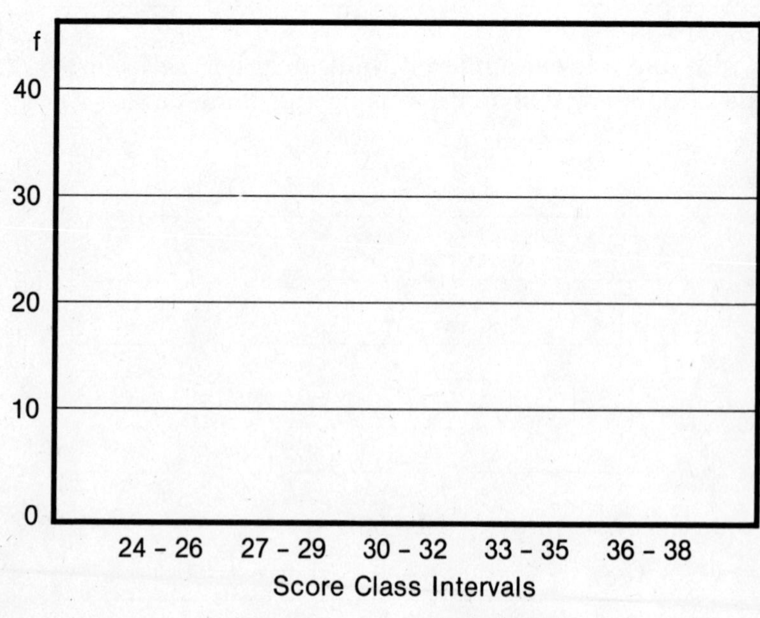

(continued)

Student ______________________________ Score ____________

SECTION 4 PRACTICE PROBLEMS (continued)

For Scoring

5. Using the results of problem 1, construct a frequency polygon. 5. _____

f

15

10

5

0

25 26 27 28 29 30 31 32 33 34 35 36

Scores

6. Scale the horizontal axis according to the scores and construct a smooth-line graph on the grid that follows. 6. _____

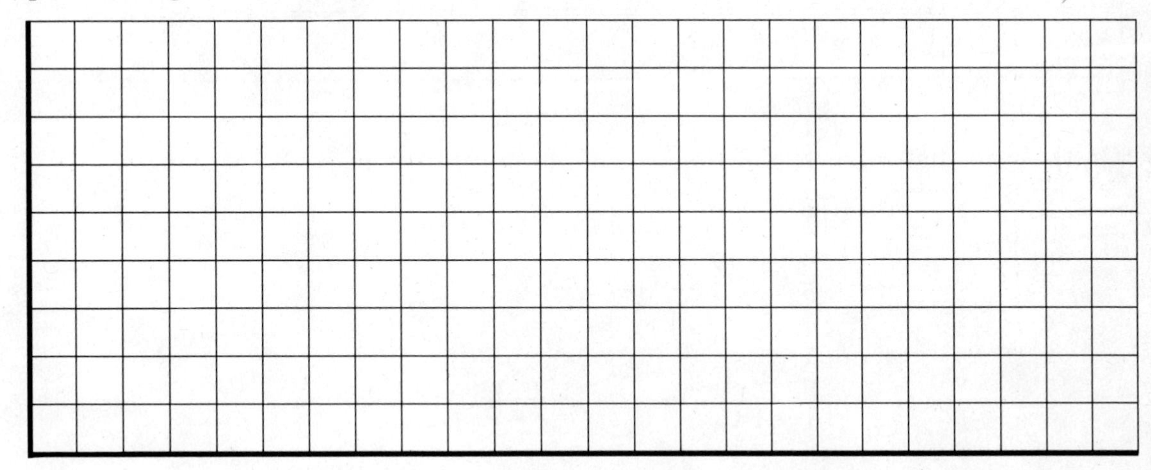

(continued)

SECTION 4 PRACTICE PROBLEMS (continued)

For Scoring

7. Use the relative cumulative frequency distribution and plot an ojive curve on the grid that follows. 7. ____

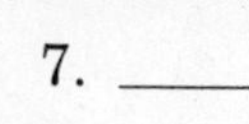

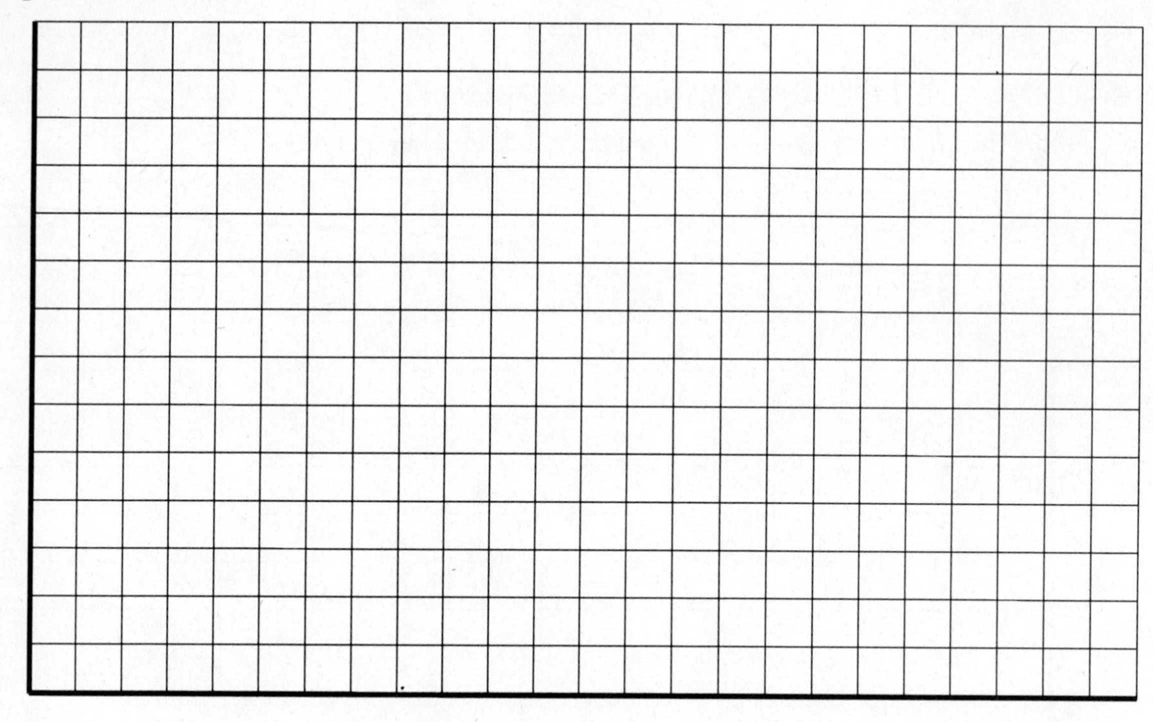

8. Complete the partially finished joint frequency distribution table that follows. 8. ____

Subject Area

	Math/Science	Social Science	Language	Voc/Tech	Marginals
Male	40	80		20	160
Female			75	10	200
Marginals	85				

9. On the grid that follows, plot the 15 ordered pairs to form a scattergram. 9. ____

X	Y	X	Y
6	8	6	9
2	3	5	5
4	6	4	7
2	4	1	3
5	7	8	9
3	6	1	2
5	8	8	8
4	5		

(continued)

Student ______________________ Score __________

SECTION 4 PRACTICE PROBLEMS (continued) **For Scoring**

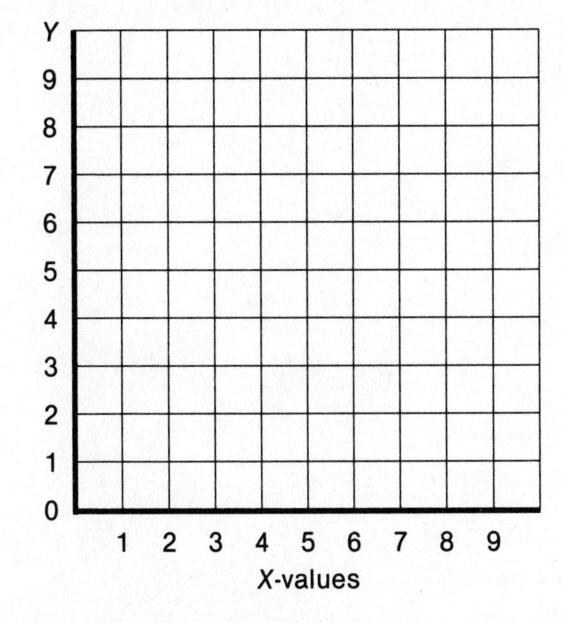

*10. Using the same data that were organized in problems 1 through 7, construct a stem-and-leaf diagram. 10. ______

Stem (10's)	Leaves (Units)
____	
____	

Following are IQ scores of a class of 50 high school students in a rural school district. Use these data for problems 11 through 14.

122	115	109	118	112
114	118	110	108	114
107	114	112	107	110
110	116	113	115	111
114	109	108	118	115
120	113	107	117	116
108	117	114	111	108
111	119	118	108	110
112	107	115	115	111
120	109	110	111	109

(continued)

For Scoring

11. Construct an ungrouped frequency distribution of the IQ scores. 11. _____

IQ	Tally	f
122	______	______
121	______	______
120	______	______
119	______	______
118	______	______
117	______	______
116	______	______
115	______	______
114	______	______
113	______	______
112	______	______
111	______	______
110	______	______
109	______	______
108	______	______
107	______	______

12. Construct a relative cumulative frequency distribution of the IQ data. 12. _____

IQ	rcf
122	______
121	______
120	______
119	______
118	______
117	______
116	______
115	______
114	______
113	______
112	______
111	______
110	______
109	______
108	______
107	______

(continued)

Student ______________________________ Score ____________

SECTION 4 PRACTICE PROBLEMS (continued) **For Scoring**

13. Display the IQ data in a histogram; scale the vertical axis as a relative frequency. 13. _____

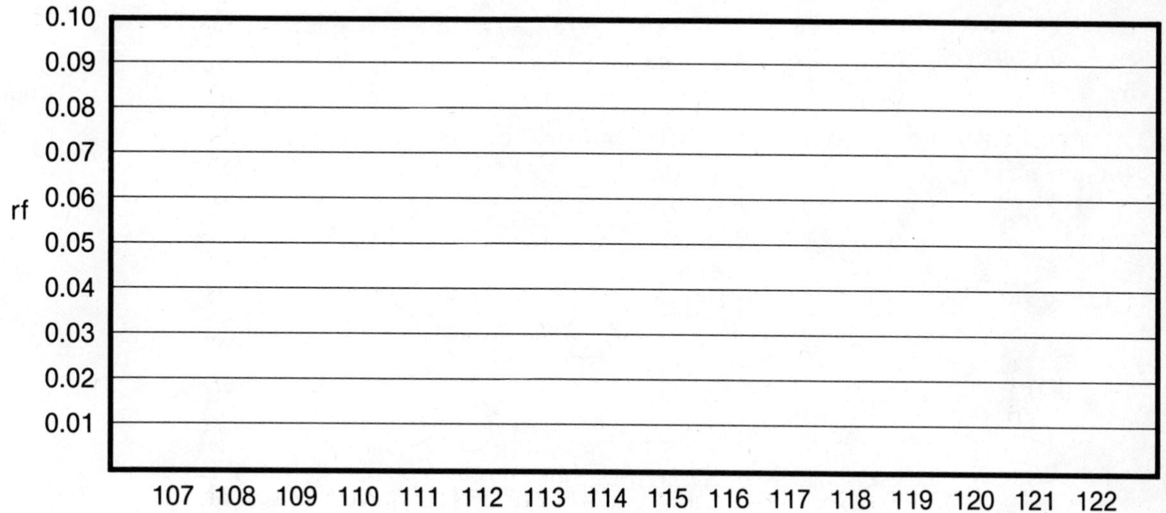

14. Using the same data, construct an ojive showing the cumulative distribution of IQ scores. 14. _____

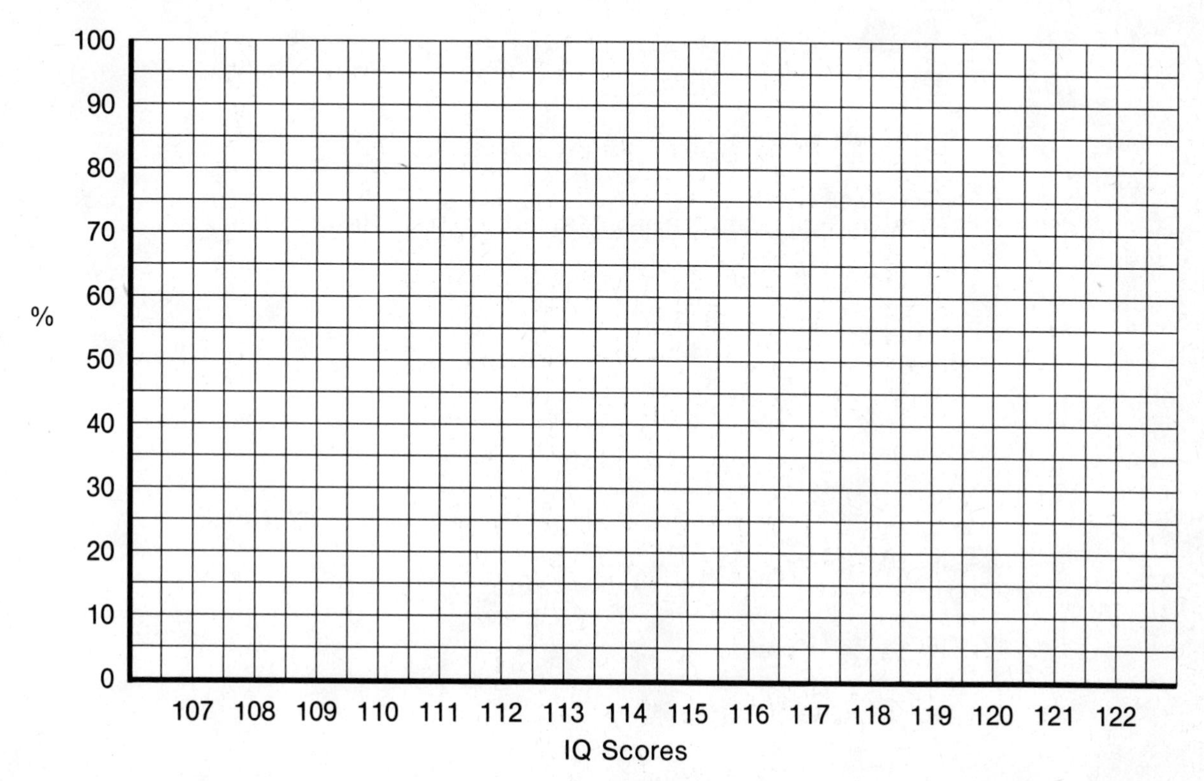

(continued)

SECTION 4 PRACTICE PROBLEMS (continued)

Following are 15 pairs of scores. The X-variable represents scores on a mathematics aptitude test, and the Y-variable represents scores on a commercial mathematics achievement test. Use this bivariate distribution in responding to problems 15 through 18.

X	Y	X	Y	X	Y
83	82	73	79	72	70
70	72	82	83	75	77
79	76	77	79	73	73
76	74	79	82	81	75
75	75	74	74	71	73

For Scoring

15. Construct the ungrouped frequency distributions for the two variables adjacent to each other. 15. _____

Score	f(X)	f(Y)
83	________	________
82	________	________
81	________	________
80	________	________
79	________	________
78	________	________
77	________	________
76	________	________
75	________	________
74	________	________
73	________	________
72	________	________
71	________	________
70	________	________

*16. Construct a stem-and-leaf diagram for each variable in the blank space that follows. 16. _____

(continued)

Student ______________________________ Score __________

SECTION 4 PRACTICE PROBLEMS (continued) **For Scoring**

17. Plot a bivariate scattergram scaling *X* horizontally and *Y* on the vertical axis. 17. _____

18. Describe the apparent trend shown in the scattergram in problem 17; that is, describe the bivariate relationship between mathematics aptitude and mathematics achievement. 18. _____

____

____

____

____

____

____

(continued)

19. Plot the bivariate distribution ($N = 13$) of X and Y on a scattergram. Then write a description of the appearance of the bivariate distribution. 19. _____

X	Y	X	Y
1	1	7	8
4	6	3	5
8	7	10	1
5	10	5	9
7	9	6	9
2	4	4	7
9	5		

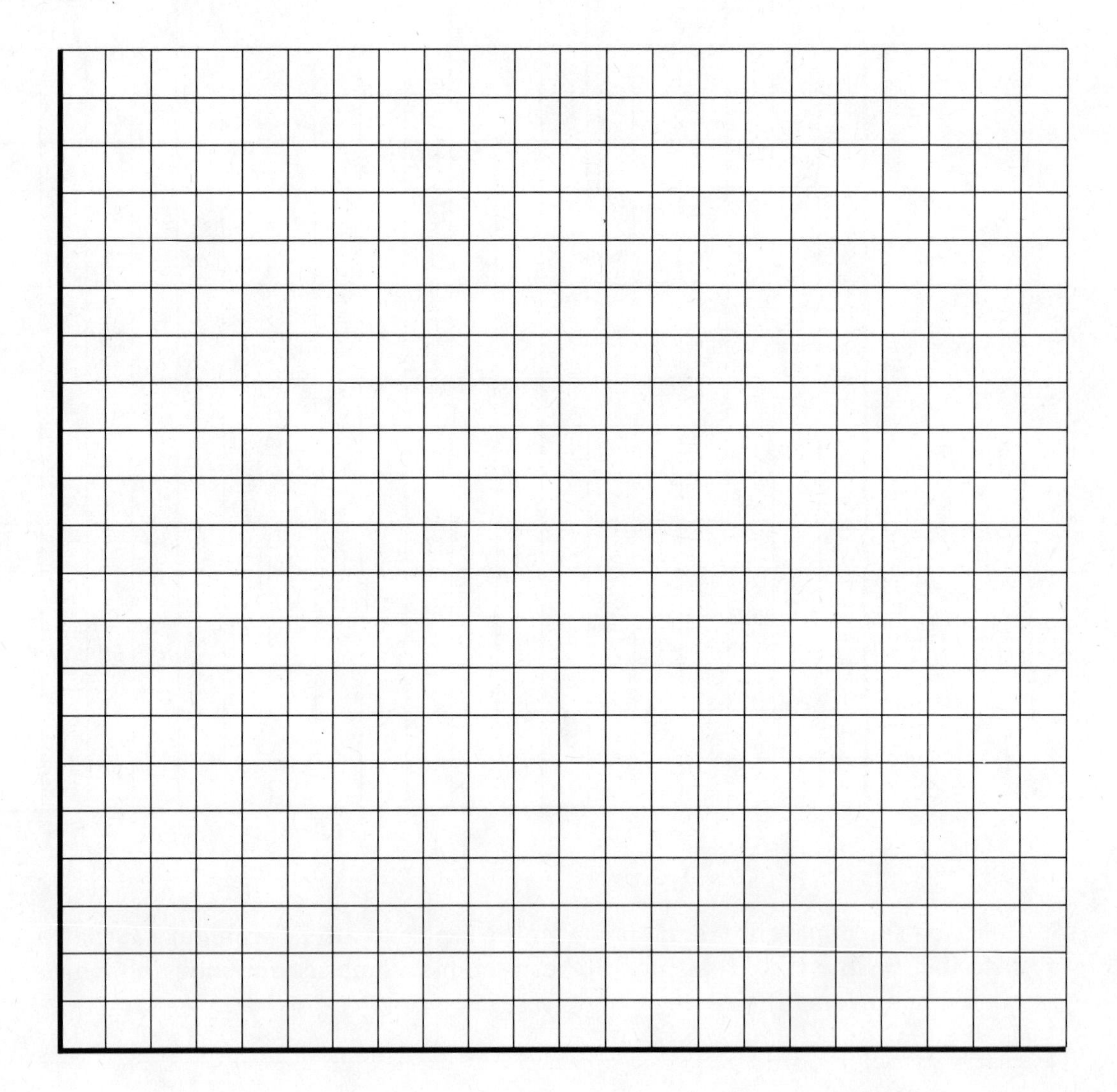

____

____

____

(continued)

Student ______________________________ Score __________

SECTION 4 PRACTICE PROBLEMS (continued)

For problems 20 through 23, refer to the descriptive analysis that follows.

One researcher reported the results of a survey of 1,250 students in a certain high school on their attitude toward corporal punishment. One question asked the students if they thought that corporal punishment in schools should be illegal. Students' responses to this question were scored from 1 (definitely should be illegal) through 5 (definitely should not be illegal), where 3 represented a neutral position. The results expressed as percentages of responses to each alternative were as follows:

Item Score	Percentage
5	18
4	22
3	10
2	31
1	19

Answers **For Scoring**

20. How many students were undecided? ______ 20. ______

21. How many students did not feel that corporal punishment should be illegal? (Hint: Find the number marking 3, 4, or 5.) ______ 21. ______

22. What percentage of the students had extreme opinions (1 or 5) about the issue? ______ 22. ______

23. Given the results of the survey, construct a table showing f, rf, cf, and rcf. 23. ______

Item Score	f	rf	cf	rcf
5	______	______	______	______
4	______	______	______	______
3	______	______	______	______
2	______	______	______	______
1	______	______	______	______

CHAPTER 4

Summarizing Data: Measures of Central Tendency

Student ______________________ Total Score ____________

SECTION 1 CONCEPT COMPREHENSION

Directions: Complete the following by writing the correct word or words in the respective blanks to the right.

Measures on data that are designed to provide information about where the central portion of a distribution lies are called __(1)__. The three most commonly used measures of central tendency are the __(2)__, __(3)__, and __(4)__. Of these, the __(5)__ mean is the one that requires all of the actual values to be used in the computation.

In some distributions the __(6)__ may not exist or may not be a unique value. The 50th percentile is also the __(7)__. The formula for the mean uses the symbol Σ, which indicates that the values are to be __(8)__. If there is an odd number of values in a particular set, the median is the __(9)__ number if they are in ascending order of size. If a distribution has two values that are tied for the most frequently occurring, the distribution is referred to as __(10)__.

If the mean, median, and mode of a distribution coincide (have the same value) and the curve to the right appears to be a reflection of the left side of the curve, the distribution is __(11)__. Such a curve is frequently called a(n) __(12)__-shaped curve. If, however, a distribution has a large number of values stacking up in the lower portion, it has a(n) __(13)__ skew. In this kind of distribution, the __(14)__ will be larger than the other two common measures of central tendency.

In addition to the arithmetic mean, this chapter describes two more means: the __*(15)__ mean and the __*(16)__ mean. One is appropriate for computing average speeds when the times traveled are different but the __*(17)__ are the same. The other is applicable to distributions in which the __*(18)__ of any two consecutive numbers is fairly constant across the distribution.

Answers	For Scoring
1. ____________	1. ____
2. ____________	2. ____
3. ____________	3. ____
4. ____________	4. ____
5. ____________	5. ____
6. ____________	6. ____
7. ____________	7. ____
8. ____________	8. ____
9. ____________	9. ____
10. ____________	10. ____
11. ____________	11. ____
12. ____________	12. ____
13. ____________	13. ____
14. ____________	14. ____
15. ____________	15. ____
16. ____________	16. ____
17. ____________	17. ____
18. ____________	18. ____

*Any item that has an asterisk before the number indicates it covers optional text material. If such material was not presented in class, you may skip this item. Check with your teacher.

SECTION 2 VOCABULARY PRACTICE

Directions: Identify the term defined and complete the designated squares.

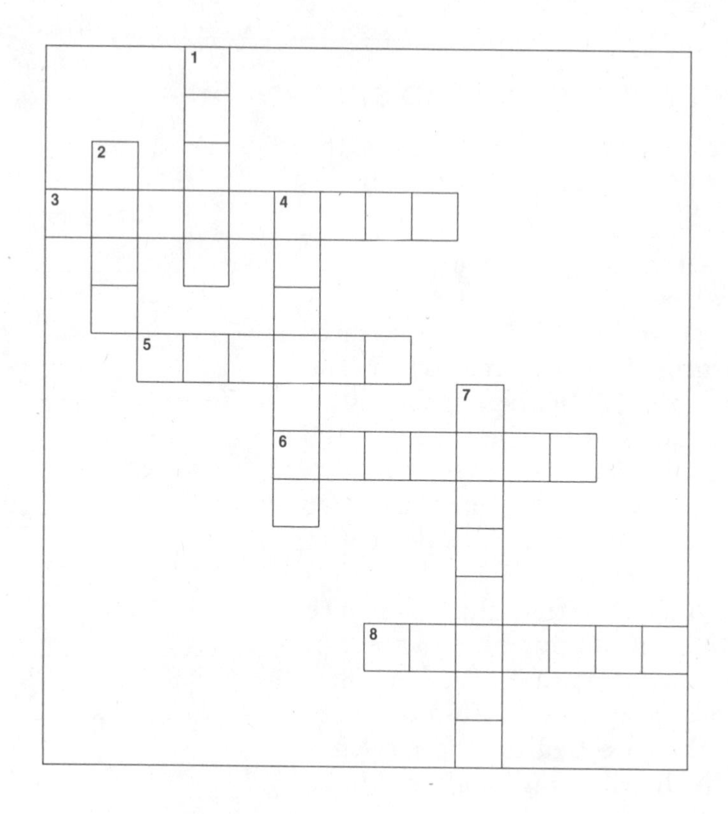

Across

For Scoring

*3. Its computation requires finding the *n*th root of a product of the numbers. ________

5. The measure of central tendency that divides a distribution so that half of the values are above and half below that point. ________

6. Another name for measure of central tendency. ________

8. The median is located in the ______ portion of a distribution. ________

Down

1. An uppercase Greek letter indicating summation. ________

2. A measure of central tendency often referred to as "*X*-bar." ________

4. An average is used to describe the ______ value in a distribution. ________

*7. A measure of central tendency frequently used to determine average speed or velocity. ________

Summarizing Data: Measures of Central Tendency

Student ______________________________ Score ____________

SECTION 3 REHEARSAL EXERCISES

Directions: In the Answers column, write the letter that represents the correct choice.

	Answers	For Scoring
1. The most widely used and important measure of central tendency is the: (A) arithmetic mean (B) sigma (C) median (D) mode	______	1. _____
2. If a group of high school biology students were given an anatomy exam developed for medical school students, it is likely that the ______ would be higher than the ______ for the biology class score distribution. (A) mode; median (B) median; mean (C) mode; mean (D) mean; median	______	2. _____
3. If the variable $X = \{2, 5, 6, 8, 9, 10\}$, what is the value of ΣX? (A) 0 (B) 40 (C) 43,200 (D) 2,568,910	______	3. _____
4. The measure of central tendency most likely to be excessively influenced by extreme values is the: (A) median (B) mode (C) arithmetic mean (D) none of the above; all are influenced the same	______	4. _____
5. Forty-two members of the varsity football team were being weighed. Unfortunately, the scale's upper limit was 220 pounds. Eight of the players weighed too much for the scales and thus could not be assessed. To get a reliable average, which of the following procedures should be used? (A) Eliminate the eight players from consideration and report the arithmetic mean of the remainder of the team. (B) Use the method in (A), except add 15 pounds to the average. (C) Determine and report the median. (D) Guess the weights for the 8 players and then find the mean of all 42, including the guesses.	______	5. _____

(continued)

SECTION 3 REHEARSAL EXERCISES (continued)

	Answers	For Scoring

6. Calculate the mean from the frequency distribution: ______ 6. ______
   (A) 2.00
   (B) 2.40
   (C) 3.40
   (D) 2.18

Variable X	f
6	2
3	2
2	6
1	4
0	3

7. If a set of test scores yielded a positively skewed distribution, more people would *fail* if the ______ was the criterion for passing than if the ______ was the criterion for passing. ______ 7. ______
   (A) mode; median
   (B) median; mean
   (C) mode; mean
   (D) mean; median

8. Which of the following is not affected by the actual magnitude of every score in the set on which it was calculated? ______ 8. ______
   (A) arithmetic mean
   (B) median
   *(C) harmonic mean
   *(D) geometric mean

9. A distribution of scores has a mean of 110, a median of 112, and a mode of 115. What score is located in the middle of the distribution (that is, the 50th percentile)? ______ 9. ______
   (A) 110
   (B) 112
   (C) 115
   (D) Cannot be determined with the information provided.

10. The distribution represented by the three measures of central tendency in problem 9 is probably: ______ 10. ______
    (A) symmetric
    (B) bimodal
    (C) negatively skewed
    (D) positively skewed

SECTION 4 PRACTICE PROBLEMS

	Answers	For Scoring

1. For the following set of numbers, find the quantities indicated: X: 59, 45, 68, 30, 57, 52, 30, 32, 65, 49.
   a. N ______ 1a. ______
   b. ΣX ______ 1b. ______
   c. $\overline{X}$ ______ 1c. ______
   d. median ______ 1d. ______
   e. mode ______ 1e. ______

(continued)

Student ______________________ Score ________

SECTION 4 PRACTICE PROBLEMS (continued) **Answers** **For Scoring**

2. For the values 16 and 25, find:
   *a. the harmonic mean ________ 2a. ____
   *b. the geometric mean ________ 2b. ____
   c. the arithmetic mean ________ 2c. ____

3. Given the following numbers: 9, 12, 16, 12, find:
   a. the arithmetic mean ________ 3a. ____
   *b. the geometric mean ________ 3b. ____
   *c. the harmonic mean ________ 3c. ____
   d. the median ________ 3d. ____
   e. the mode ________ 3e. ____

4. Suppose six cards are tossed on a table 100 times. The number of cards falling face up (X) and the respective frequencies (f) were recorded as follows:

X:	0	1	2	3	4	5	6
f:	1	11	24	34	23	5	2

   a. Find the mean of the X-distribution. ________ 4a. ____
   b. Find the median of the X-distribution. ________ 4b. ____
   c. What is the mode? ________ 4c. ____

*5. The approximate population of the United States from 1960 to 1990 is as follows: ________ 5. ____

Year	Population
1960	179.3
1970	203.3
1980	226.5
1990	254.7

Compute the geometric mean of the population for that time period.

*6. A family took two trips to San Diego. On the first trip they averaged 40 mph to San Diego and 50 mph on their return. For the second trip they averaged 53 mph to San Diego and 48 mph when returning. What was their overall average speed for both trips? ________ 6. ____

(continued)

For Scoring

7. Tell what is wrong with the measures of central tendency used for the following purposes.
   a. Coach Sampson is trying to determine what size athletic shoes in various quantities should be ordered for the athletic department. The coach examines the records for the previous three years and then calculates the arithmetic mean to determine the size most frequently required (worn).

   7a. ______

   b. Bobby wanted to impress his parents with the allowances his four best friends were getting. He presented his parents with the mean allowance. Bobby received $10 per week. His four friends were getting $8, $9, $10, and $35 per week, respectively.

   7b. ______

   c. To determine the most frequently occurring class size in a school, the principal selected the median class size.

   7c. ______

   *d. The average speed of a school bus during its morning route was 30 mph, and in the afternoon the average speed was 23 mph for the same route. The arithmetic mean was used to determine the overall average speed.

   7d. ______

(continued)

Summarizing Data: Measures of Central Tendency

Student ______________________________ Score __________

SECTION 4 PRACTICE PROBLEMS (continued)

Answers | For Scoring

8. A survey was conducted in a school system to determine how many consecutive years individual teachers have been assigned to the same home room in their building. The frequency analysis for these data is as follows:

Number of Years	f
10	12
9	16
8	10
7	15
6	8
5	13
4	13
3	10
2	9
1	16

a. Compute the mean. __________ 8a. ______

b. Compute the median. __________ 8b. ______

c. Determine the mode(s). __________ 8c. ______

9. Suppose you record the weights of 50 freshman students at your school's "Fitness Day." However, you learn that the calibration of your scales had made the recorded weights for each individual too heavy by one pound. How would you correct the average weight of the freshmen? 9. ______

____

____

____

10. In a survey of members of the high-school statistics class, the distribution of scores on the midterm exam was as follows: 10. ______

Score	f
90 or higher	3
88	2
85	3
84	4
83	3
81	2
79	1
78	2
75	3
less than 75	2

(continued)

 For Scoring

You have been commissioned to find the "average" or "typical" score on the midterm exam. Perform this task and justify your methods.

____

____

____

____

11. If most of your classmates in statistics had studied Chapter 4 so carefully that they knew the answers to all or almost all of the questions in these exercises, the distribution of scores for this activity would probably take on a shape described as ______________. In such a case, describe the relationship among the three measures of central tendency. 11. ____

____

____

12. Six statistics students live along a straight stretch of highway from their school. Their homes are positioned along the highway as follows:

School	4 mi	Dave	2 mi	Sue	3 mi	Emilia	1 mi	Juan	2 mi	Eva	1 mi	Ed

Because any of the seven locations is suitable for a group study session, the students want to meet at a location such that the total amount of travel will be minimized.

Answers

a. Which location would satisfy the requirement? ________ 12a. ____

b. Considering the distance from the school to the six locations as a variable, which measure of central tendency corresponds to your answer to (a)? (This illustrates that the sum of the absolute values of the deviation scores from this measure of central tendency is a minimum.) ________ 12b. ____

13. A professor at a local university teaches a morning class of $N = 56$ students and an afternoon section of the same subject of $N = 78$ students. On the first exam of the semester, the mean score for the morning group was 84.8 and the mean for the afternoon group was 80.3. The overall mean for the combined classes would be the mean of the entire group ($N = 134$) of students. Use the definition of μ to determine the overall mean. ________ 13. ____

14. A science class performed an experiment in which reaction time was assessed. The following reaction times measured in seconds were recorded: ________ 14. ____

1.2	0.9	1.1	1.0	0.9	0.8	0.6	1.1
1.0	0.8	0.7	1.0	0.8	1.1	1.0	1.2

Compute the mean for these reaction times. (Hint: First multiply each time by 10 to get rid of fractions; second, perform the calculations to find the mean; then divide the result by 10 to return to the correct time.)

Summarizing Data: Measures of Dispersion

Student ______________________________ Total Score __________

SECTION 1 CONCEPT COMPREHENSION

Directions: Complete the following by writing the correct word or words in the respective blanks to the right.

Scatter, spread, and variability are terms used to describe the degree of variation in a set of data. Just as with central tendency, several measures of __(1)__ are used to quantify this characteristic of a distribution. The __(2)__ is a simple one to determine because it indicates the spread from the lowest value to the highest value in the distribution. The average amount by which the values deviate from the distribution arithmetic mean is called the __(3)__ deviation.

When the mean is subtracted from a value in the set, that difference is known as a(n) __(4)__ score and is symbolized with a(n) __(5)__. The sum of these differences always yields a value of __(6)__ within rounding tolerances. However, when these differences are squared and added, an important ingredient of variability known as the __(7)__ of the deviation scores is obtained.

One important measure of variability known as the __(8)__ is obtained by dividing Σx^2 by N. Of more use for interpretation purposes, the square root of the variance is a value called the __(9)__ deviation and is designated by the Greek letter __(10)__.

	Answers	For Scoring
1.	______	1. ___
2.	______	2. ___
3.	______	3. ___
4.	______	4. ___
5.	______	5. ___
6.	______	6. ___
7.	______	7. ___
8.	______	8. ___
9.	______	9. ___
10.	______	10. ___

SECTION 2 VOCABULARY PRACTICE

Directions: Identify the term defined and complete the designated squares.

Across

For Scoring

1. The characteristic of spread or scatter of a numerical distribution. ________
3. The sum of squares divided by N. ________
4. The ______ deviation provides an indication of how far, on the average, the values of a distribution are from the arithmetic mean. ________
6. The ______ deviation is an important measure of dispersion that involves finding the square root of a value. ________
7. The variance is the ______ of the standard deviation. ________

Down

1. A ______ score is the algebraic difference between a raw score and the distribution mean. ________
2. The distance between the high and low values in a set. ________
5. The difference between two values when the algebraic sign is ignored is the ______ difference. ________
7. The meaning of the symbol sigma (Σ). ________

Summarizing Data: Measures of Dispersion

Student ______________________________ Score __________

SECTION 3 REHEARSAL EXERCISES

Directions: In the Answers column, write the letter that represents the correct choice.

	Answers	For Scoring
1. Which of the following *does not* conceptually belong with the other three? (A) mean (B) range (C) median (D) average	______	1. ______
2. Analogy: "Arithmetic mean" is to "central tendency" as "standard deviation" is to ______. (A) average (B) variability (C) median (D) range	______	2. ______
3. Two classes took the same exam. Class M's scores were distributed with a standard deviation of 6.4, and class Q's standard deviation was 4.0. Which of the following statements is true? (A) Class M had a higher average score than class Q. (B) The variance of the scores of class Q is 2. (C) Class Q's score distribution was more homogeneous than class M's distribution. (D) Class M's distribution was 16 times as dispersed as class Q's distribution.	______	3. ______
4. Which of the following *is not* a measure of dispersion? (A) range (B) standard deviation (C) variance (D) sum of the deviation scores	______	4. ______
5. One standardized test is distributed with $\overline{X} = 100$ and $\sigma = 25$. What is the variance of the test? (A) 625 (B) 25 (C) 5 (D) 10	______	5. ______
6. Given the following set of scores: 2, 6, 4, 3, 5, how far, on the average, do the scores deviate from the mean? (A) 1.2 points (B) 4.0 points (C) 0 points (D) 6.0 points	______	6. ______

(continued)

SECTION 3 REHEARSAL EXERCISES (continued)

	Answers	For Scoring
7. The quantity Σx^2 is equal to: (A) ΣX^2 (B) $\Sigma(X - \overline{X})$ (C) $(\Sigma X)^2$ (D) $\Sigma X^2 - \frac{(\Sigma X)^2}{N}$	______	7. ______
8. Given that $\Sigma X = 693$ and $N = 12$, the variance is: (A) 57.75 (B) 3,335.06 (C) 91.19 (D) Cannot be determined from the information provided.	______	8. ______
9. The absolute value of +36 is: (A) −36 (B) 36 (C) 6 (D) 1,296	______	9. ______
10. For any numerical distribution, the algebraic sign of the expression $\Sigma(X - \overline{X})$ will be: (A) positive (B) negative (C) could be either positive or negative (D) neither positive nor negative	______	10. ______

SECTION 4 PRACTICE PROBLEMS

Use the following collection of scores in problems 1 through 12:

0, 2, 4, 16, 8, 10, 12, 14, 18, 16

	Answers	For Scoring
1. Calculate the median.	______	1. ______
2. Calculate the arithmetic mean.	______	2. ______
3. Find the range.	______	3. ______
4. Compute the mean (average) deviation.	______	4. ______
5. Calculate the sum of squares.	______	5. ______
6. Compute the variance.	______	6. ______
7. Find the standard deviation.	______	7. ______
8. If 6 was added to each of the scores in the collection, what would be the standard deviation of the new distribution?	______	8. ______
9. If each score was multiplied by 2, what would be the new standard deviation?	______	9. ______
10. What is the value of $\Sigma(X - 10)$?	______	10. ______
11. What is the value of x (the deviation score) that corresponds to a raw score of 6?	______	11. ______
12. What is the mode of the set of numbers?	______	12. ______

(continued)

Student ______________________ Score __________

SECTION 4 PRACTICE PROBLEMS (continued)

Answers **For Scoring**

13. Use summary statistics to describe the following set of scores on the unit exam for 20 students in a civics class.

92	86	82	84	93	80	71	75	88	73
74	62	81	98	94	82	80	80	74	79

Central Tendency

a. mean ______ 13a. ______

b. median ______ 13b. ______

c. mode ______ 13c. ______

Dispersion

d. range ______ 13d. ______

e. variance ______ 13e. ______

f. standard deviation ______ 13f. ______

g. mean deviation ______ 13g. ______

14. Two sets of data are:

Set A: 4, 5, 10, 8, 7, 7, 4, 6, 7, 8, 5, 6
Set B: 4, 10, 8, 8, 7, 5, 6, 8, 8, 9, 8, 8, 8

a. How do the ranges of the two data sets compare? 14a. ______

____

b. How do the mean deviations compare? 14b. ______

____

c. Which has the larger variance? 14c. ______

____

d. Sketch histograms for the two sets of data on the same axes that follow and visually inspect the variation summarized quantitatively in a., b., and c.

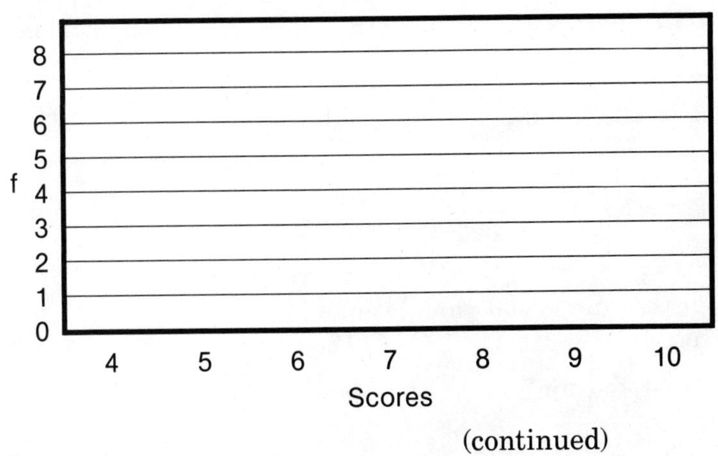

(continued)

Answers | For Scoring

15. When a data set consists of only two scores, the standard deviation becomes very easy to determine. For example, consider a set in which $X = 3$ and $X = 7$. Notice that the mean is halfway between the scores; that is, $\mu = 5$. Each score deviates two units from μ. And, if every score has a deviation of two units, then σ must be equal to 2. Let's see.

$$\begin{aligned}\Sigma x^2 &= \Sigma X^2 - \frac{(\Sigma X)^2}{N} \\ &= 58 - \frac{(10)(10)}{2} \\ &= 58 - 50 = 8\end{aligned}$$

and $\sigma = \sqrt{\frac{8}{2}} = \sqrt{4} = 2$.

Use this general rule (instead of the computational formula) to find the standard deviation of each of the following sets of data with $N = 2$:

a. $X = 14, 28$ ______ 15a. ______

b. $X = 5, 10$ ______ 15b. ______

c. $X = 0, 1$ ______ 15c. ______

d. $X = 93, 115$ ______ 15d. ______

16. Compute the standard deviations for the two sets of data shown.

Data set A: 6, 4, 5, 6, 4, 3, 5
Data set B: 6, 4, 5, 6, 4, 3, 17

a. Standard deviation of set A is ______. ______ 16a. ______

b. Standard deviation of set B is ______. ______ 16b. ______

c. What is the primary contributor to the discrepancy in the two standard deviations? 16c. ______

______

17. How do you think the following factors would tend to affect the *range* of a distribution; that is, would the factor increase or decrease the range, or would it remain essentially unchanged?

a. Include several additional extreme scores. 17a. ______

______

b. Increase N (number of observations or measurements). 17b. ______

______

c. Remove several values close to the middle of the distribution. 17c. ______

______

d. Add 10 to each value in the distribution. 17d. ______

______

18. Without actually going through the numerical calculations,

a. Find Σx^2 for this set of scores: 12, 12, 12, 12, 12, 12, 12. ______ 18a. ______

b. What is the variance of the distribution? ______ 18b. ______

c. What is the mean deviation? ______ 18c. ______

CHAPTER 6

Describing Individual Performances

Student ______________________________ Total Score ____________

SECTION 1 CONCEPT COMPREHENSION

Directions: Complete the following by writing the correct word or words in the respective blanks to the right.

__(1)__ scores are values that have been collected without transformation or alteration. The scores can be computationally transformed to a standard *z*-score scale if the __(2)__ and __(3)__ deviation are known as measures of central tendency and dispersion, respectively. A distribution of *z*-scores then has a mean of __(4)__ and a standard deviation of __(5)__. A particular *z*-score is a point on the baseline indicating how many __(6)__ the point is from the mean. A negative *z*-score means the corresponding raw score is __(7)__ the distribution mean.

If each *z*-score in a distribution is multiplied by 10 and the products are then increased by 50, the new distribution consists of scores called __(8)__. This distribution has a mean of __(9)__ and a standard deviation of __(10)__. Further, with the help of a table of normal curve values, *z*-scores can be changed to another commonly used score distribution called __(11)__. While these scores are not standard scores, they indicate the percentages of values equal to or __(12)__ a particular score. Relative to quartiles, the 25th, 50th, and 75th percentile ranks are designated as __*(13)__, __*(14)__, and __*(15)__, respectively.

The 25th and 75th percentiles serve as __*(16)__ on a box-and-whisker plot. The difference between these two points is called the __*(17)__ range. In this diagram, values that are extreme and are beyond the adjacent values are referred to as __*(18)__.

Answers	For Scoring
1. ____________	1. ____
2. ____________	2. ____
3. ____________	3. ____
4. ____________	4. ____
5. ____________	5. ____
6. ____________	6. ____
7. ____________	7. ____
8. ____________	8. ____
9. ____________	9. ____
10. ____________	10. ____
11. ____________	11. ____
12. ____________	12. ____
13. ____________	13. ____
14. ____________	14. ____
15. ____________	15. ____
16. ____________	16. ____
17. ____________	17. ____
18. ____________	18. ____

*Any item that has an asterisk before the number indicates it covers optional text material. If such material was not presented in class, you may skip this item. Check with your teacher.

SECTION 2 VOCABULARY PRACTICE

Directions: Identify the term defined and complete the designated squares.

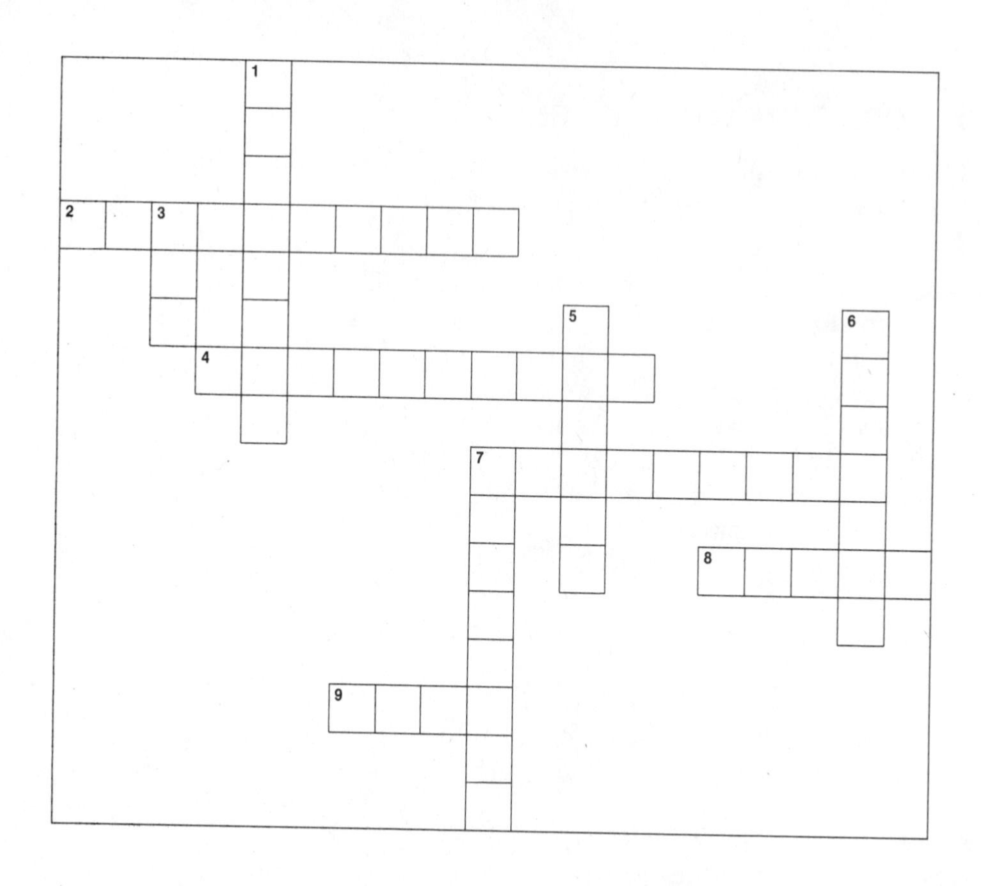

Across

For Scoring

2. A score that indicates the percentage of values at or below a particular value. ________
4. A point on a curve at which the curve changes from concave upward to concave downward or vice versa. ________
7. A ______ curve is the same on both sides of the mean. ________
8. A smooth-line graph of a frequency distribution. ________
9. Half of the ______ under the curve (between the curve and the baseline) is to the left of the 50th percentile. ________

Down

1. *z*-scores are scaled along the ______ in standard deviation units. ________
3. Scores that have not been transformed to standard scores or percentiles. ________
5. A theoretical distribution that closely approximates many real data frequency distributions. ________
6. A term used to indicate that a curve is bending toward or away from the baseline. ________
7. *T*-scores and *z*-scores are more commonly called ______ scores. ________

Describing Individual Performances

Student ______________________________ Score __________

SECTION 3 REHEARSAL EXERCISES

Directions: In the Answers column, write the letter that represents the correct choice.

Answers For Scoring

1. A normal distribution of scores has a mean of 92 and a standard deviation of 8. If you had a score of 104, what would be the corresponding z-score? ______ 1. ______
   (A) 0.93
   (B) 1.50
   (C) 0.79
   (D) 65

2. A normal distribution has a mode of 75 and a median of.75. The standard deviation is 5. A score of 65 in this distribution could be represented by: ______ 2. ______
   (A) $z = -2.00$
   (B) T-score = 30
   (C) about the 2.3rd percentile
   (D) All of the above are correct.

3. A normal distribution has $\overline{X} = 20$ and $\sigma = 5$. A score of 30, when transformed to a distribution with a mean of 100 and a standard deviation of 15, would be: ______ 3. ______
   (A) 115
   (B) 95
   (C) 130
   (D) none of the above

4. Sally missed 12% of the questions on a biology test. This means that: ______ 4. ______
   (A) Her score is the 12th percentile.
   (B) Her score is the 88th percentile.
   (C) She received a T-score of 62.
   (D) None of the above is necessarily true.

5. Assume that a distribution of raw scores on an examination has been changed to standard z-scores. This z-score distribution is then multiplied by 30, yielding a new distribution with a mean of ______ and a standard deviation of ______. ______ 5. ______
   (A) 30; 1
   (B) 0; 30
   (C) 30; 30
   (D) 1; 30

(continued)

SECTION 3 REHEARSAL EXERCISES (continued)

Answers | For Scoring

6. The mean scores and standard deviations for final examinations in algebra were $\overline{X} = 60$ and $\sigma = 12$; in statistics, $\overline{X} = 58$ and $\sigma = 5$; and in physics, $\overline{X} = 56$ and $\sigma = 7$. Neil scored 72 on the algebra final, Don scored 68 on the statistics final and Juanita scored 77 on the physics final. How did the three students perform relative to the three respective course exams? ______ 6. ______
   (A) Neil best; Don second; Juanita third
   (B) Juanita best; Don second; Neil third
   (C) Juanita best; Neil second; Don third
   (D) Neil best; Juanita second; Don third

7. A student has a z-score of -0.50. What is the value of the corresponding T-score? ______ 7. ______
   (A) 45.0
   (B) 50.0
   (C) 55.0
   (D) none of the above

8. The 30th percentile is the point in a distribution of scores: ______ 8. ______
   (A) that marks the distance from the mean such that 30% of the scores are included between that point and the mean
   (B) where a student has answered 30% of the questions correctly
   (C) below which are 30% of the cases
   (D) that corresponds to a T-score of 30

*9. The percentage of cases represented between the hinges of a box-and-whisker plot is: ______ 9. ______
   (A) 68%
   (B) 34%
   (C) 50%
   (D) 100%

*10. If the median is markedly to the left of the center of the rectangle in a box-and-whisker diagram, probably the distribution is: ______ 10. ______
   (A) positively skewed
   (B) negatively skewed
   (C) symmetric
   (D) normal (bell shaped)

SECTION 4 PRACTICE PROBLEMS

For Scoring

1. Given the following collection of scores: 1, 4, 4, 5, 5, 6, 7, 8
   a. Calculate the mean and standard deviation. 1a. ______

      Mean = ______; σ = ______

   b. Calculate a z-score for each value in the collection; then transform all of the scores into a new distribution with a mean of 50 and a standard deviation of 10. 1b. ______

(continued)

Student ______________________ Score ______

SECTION 4 PRACTICE PROBLEMS (continued) | **Answers** | **For Scoring**

Raw Score	*z*-Score	*T*-Score
1	______	______
4	______	______
5	______	______
6	______	______
7	______	______
8	______	______

2. Given a score point located 1.65 standard deviations above the mean in a normal distribution, determine:
   a. the proportion of the area between the mean and the score ______ 2a. ______
   b. the proportion of the area to the left of the score ______ 2b. ______
   c. the proportion of the area to the right of the score ______ 2c. ______
   d. the percentile equivalent of the score ______ 2d. ______

3. Given a score point located 1.65 standard deviations below the mean in a normal distribution, determine:
   a. the proportion of the area between the mean and the score ______ 3a. ______
   b. the proportion of the area to the left of the score ______ 3b. ______
   c. the proportion of the area to the right of the score ______ 3c. ______
   d. the percentile equivalent of the score ______ 3d. ______

4. A percentile rank of 20 in a normal distribution corresponds to a *z*-score of ______. ______ 4. ______

5. A percentile rank of 80 in a normal distribution corresponds to a *z*-score of ______. ______ 5. ______

6. Given a percentile of 50 in a normal distribution, find the corresponding *z*-score. ______ 6. ______

7. What proportion of the scores in a normal distribution is higher than a *z*-score of 1.96? ______ 7. ______

8. If you had access to a set of standard *z*-scores and wished to identify the particular *z*-score that would serve as the lower bound for the top 10% in the group, what is the value of the *z*-score required? ______ 8. ______

(continued)

	Answers	For Scoring
9. Assuming the same data as in problem 8, what *z*-score range would include the middle 50% of the distribution?	______	9. ____
10. If the bottom 15% of a class received a grade of F, what *z*-score would serve as the lower bound for receiving a passing grade?	______	10. ____
11. Find the values called for and transform the following raw score values to the appropriate standard scores. Then use the normal curve table to determine the corresponding percentiles.		11. ____

	X	*z*-Score	*T*-Score	Percentile
$N =$ ______	16	______	______	______
$\overline{X} =$ ______	22	______	______	______
$(\Sigma X)^2 =$ ______	31	______	______	______
$\Sigma X^2 =$ ______	19	______	______	______
$\Sigma x^2 =$ ______	16	______	______	______
$\sigma^2 =$ ______	25	______	______	______
$\sigma =$ ______	22	______	______	______
	28	______	______	______
	25	______	______	______
	34	______	______	______
	28	______	______	______
	34	______	______	______

*12. Answer problems a. through h. given the following set of data. (Hint: Set up a frequency distribution and determine points that represent dividing points for the bottom quarter, the middle, and the top quarter of the distribution.)

17	15	20	14	17	17	13	16	18	15
19	21	19	16	15	16	17	11	17	16
12	16	17	10	18	14	21	17	19	18
17	18	22	17	18	23	19	15	17	16
18	13	17	20	14	16	20	18	24	17

	Answers	For Scoring
a. The value of Q_3 is ______.	______	12a. ____
b. The value of Q_2 (median) is ______.	______	12b. ____
c. The value of Q_1 is ______.	______	12c. ____
d. Therefore the hinges for a box-and-whisker plot are ______ and ______.	______	12d. ____
e. What is *I*, the interquartile range?	______	12e. ____
f. When *I* is multiplied by 1.5, the resulting value is used to determine the adjacent values, or ______, of the box-and-whisker plot.	______	12f. ____

(continued)

Student ______________________ Score ________

SECTION 4 PRACTICE PROBLEMS (continued) | **Answers** | **For Scoring**

g. For this distribution, the upper outliers would be located above the value of ______ on a box-and-whisker diagram. ________ 12g. ____

h. Similarly, the lower-level outliers would fall below the value of ______. ________ 12h. ____

*13. Assume that these data approximate a normal distribution with $\overline{X} = 17$ and $\sigma = 2.77$. Then use the normal curve table to determine:

a. Q_1 (25th percentile) ________ 13a. ____

b. Q_2 (median or 50th percentile) ________ 13b. ____

c. Q_3 (75th percentile) ________ 13c. ____

d. How do the values of the hinges using the normal curve approximation compare to the results in problem 12? 13d. ____

____

____

____

14. Twelve students were given a 60-item exam on civil liberties. Their scores were: 29, 53, 47, 17, 35, 23, 41, 53, 35, 41, 17, and 29.

a. Find the mean of the distribution of exam scores. ________ 14a. ____

b. Compute the standard deviation of the scores. ________ 14b. ____

c. What *z*-score corresponds to a raw score of 41? ________ 14c. ____

d. What *z*-score corresponds to a raw score of 35? ________ 14d. ____

e. A *z*-score of -1.0 corresponds to what raw score value? ________ 14e. ____

f. Suppose a distribution of standard scores is defined with a mean of 100 and a standard deviation of 10. On such a scale, what new standard score value would correspond to a raw score of 53? ________ 14f. ____

g. On the distribution defined in f. (mean = 100, standard deviation = 10), a score value of 85 would correspond to what *z*-score? ________ 14g. ____

h. The raw score on the exam corresponding to the transformed score of 85 defined in g. would be ______. ________ 14h. ____

i. A raw score of 17 on the exam, if transformed to a scale with a mean of 100 and a standard deviation of 10, would be equal to ______. ________ 14i. ____

(continued)

	Answers	For Scoring
15. An educational testing organization has designed a standardized test of statistical aptitude. Scores on this test are normally distributed with a mean of 500 and a standard deviation of 80 for high school students.		
a. What is the proportion of high school students' test scores between 500 and 600?	________	15a. ____
b. What is the proportion of scores between 460 and 540?	________	15b. ____
c. What proportion of the exam scores will exceed 350?	________	15c. ____
d. What proportion of the exam scores will fall below 475?	________	15d. ____
e. What proportion of the scores will fall between 400 and 520?	________	15e. ____
f. The 70th percentile on the test will correspond to what score on the exam?	________	15f. ____
16. In one rural county, the conviction rate for speeding by high school students is 90%. There are 200 such speeding summonses issued each year. In what proportion of the years will the number of convictions be 175 or more? ($\sigma = 3.85$)	________	16. ____
17. At Del Norte High School, the time required for students to pass from one class to another during passing period is normally distributed with a mean of 400 seconds and a standard deviation of 50 seconds.		
a. What proportion of the passing periods is completed between 288 and 400 seconds?	________	17a. ____
b. What proportion of the passing periods requires more than 492 seconds?	________	17b. ____
c. What is the proportion of passing periods requiring between 390 and 480 seconds?	________	17c. ____
d. What proportion of the passing periods requires between 360 and 440 seconds?	________	17d. ____
e. Of the 1,300 passing periods during the academic year, how many will require longer than 260 seconds?	________	17e. ____
f. Eighty percent of the passing periods will be completed in less than how many seconds?	________	17f. ____
g. The longest 5% of the passing periods require at least how many seconds?	________	17g. ____
h. What proportion of the passing periods is less than 350 or greater than 450 seconds?	________	17h. ____

For problems 18 through 20, use the following frequency distribution, which displays data on the number of items correctly answered on a 10-item true-false test by a certain high-school class.

Number Correct	f	Number Correct	f
10	2	6	7
9	3	5	7
8	8	4	2
7	9	3	1

(continued)

Student ______________________ Score ________

SECTION 4 PRACTICE PROBLEMS (continued) | **Answers** | **For Scoring**

18. a. Determine the distribution mean. ________ 18a. ____

b. Find the standard deviation. ________ 18b. ____

19. Convert each score in the distribution to a standard *z*-score and to a standard *T*-score. 19. ____

Number Correct	*z*-Score	*T*-Score
10	________	________
9	________	________
8	________	________
7	________	________
6	________	________
5	________	________
4	________	________
3	________	________

20. Using the theoretical normal curve values from Appendix C, convert each of the raw scores from problem 19 to a percentile equivalent. 20. ____

Number Correct	Percentile
10	________
9	________
8	________
7	________
6	________
5	________
4	________
3	________

21. If Penny obtained a score of 80 on a biology test, which one of the following distributions would yield the most favorable (highest) interpretation of her score? Why? 21. ____

(A) mean = 70; standard deviation = 5
(B) mean = 70; standard deviation = 15
(C) mean = 70; standard deviation = 25
(D) mean = 70; standard deviation = 40

____

____

(continued)

 Answers **For Scoring**

22. Suppose Jennifer obtained the following percentile scores on four subtests of an abilities test battery:

Subtest	Percentile
Language Skills	98
Mathematics Computation	68
Reference Usage	90
Motor Coordination	34

a. If Jennifer improved her motor coordination by one standard deviation, her new percentile equivalent score would be ______. ______ 22a. ______

b. How many standard deviations above the mean is she scoring on Language Skills? ______ 22b. ______

c. What is her z-score in Language Skills? ______ 22c. ______

d. If her score in Math Computation was decreased by 1 σ, what would her new T-score equivalent be? ______ 22d. ______

23. For a reading test, the manual states that the mean score of beginning sixth-grade students is 6.0 (these units are referred to as "grade equivalents"). Assuming a normal distribution and $\sigma = 1.6$, respond to the following items:

a. What percentage of the students entering sixth grade score below 6.0? ______ 23a. ______

b. What percentage score below a grade equivalent of 5.0? ______ 23b. ______

c. What percentage score between 6.0 and 7.2? ______ 23c. ______

d. If a state provides additional funds ($200 per student) for students scoring three or more years below grade level in reading, how much additional money would you expect would be required for a typical school district with 2,500 sixth-grade students on the basis of reading scores of the sixth-grade students? ______ 23d. ______

24. Assume that the mean of a distribution of Y-scores is 14 and the standard deviation of the Y-distribution is 4. Find the Y-score values corresponding to each of the following z-scores:

a. 1.40 ______ 24a. ______

b. 2.05 ______ 24b. ______

c. -0.25 ______ 24c. ______

d. -1.68 ______ 24d. ______

Elementary Probability

Student ______________________________ Total Score ____________

SECTION 1 CONCEPT COMPREHENSION

Directions: Complete the following by writing the correct word or words in the respective blanks to the right.

Following are some questions that refer to the seven rules of simple probability described in Chapter 7 of the textbook. Assume that some random drawings for various prizes are to be held at the annual school carnival. Twenty ($N = 20$) ping-pong balls are to be used for some of the drawings. The ping-pong balls have been painted: five balls have been painted red, eight are blue, and the remaining seven are yellow. In addition to the balls, a large box contains 20 blocks that have been numbered 1 through 20 and have their respective numbers written on the surface.

As an illustration of rule 1, suppose that blindly drawing the block with the number 13 on it from the box results in the awarding of two free passes to a movie theater. On a single trial, the probability of a winning number being drawn is __(1)__. Drawing a red ball in a single trial results in the awarding of a school banner. On any particular trial, using rule 3 we find that the probability of winning a banner is __(2)__. The meaning of rule 2 is fairly obvious and entails a minimum and maximum likelihood of certain events occurring. For example, the probability of drawing a green ping-pong ball is __(3)__, while the probability of selecting a colored ball on a single trial is __(4)__. Or, with rule 4, the problem could be posed: What is the probability of selecting a red or a blue or a yellow ping-pong ball? In this case the solution is __(5)__, the same as the previous answer.

Given that the probability of selecting a red ping-pong ball at random is 0.25, rule 5 implies that the probability of selecting a ball that is some other color than red in a single trial is __(6)__. On a single trial, the probability of drawing a block with a 3, 6, 9, or 12 on it could be symbolized as $p(3, 6, 9, 12)$. Because the individual probability that each of the numbers would occur in a single trial is $\frac{1}{20}$, rule 6 implies that $p(3, 6, 9, 12) =$ __(7)__. Rule 7, sometimes known as the __(8)__ rule, applies to joint events. If a ping-pong ball is drawn and a block is selected at the same time, the probability that the ball is yellow *and* the number on the block is an even number is __(9)__.

Answers	For Scoring
1. ________	1. ____
2. ________	2. ____
3. ________	3. ____
4. ________	4. ____
5. ________	5. ____
6. ________	6. ____
7. ________	7. ____
8. ________	8. ____
9. ________	9. ____

SECTION 2 VOCABULARY PRACTICE

Directions: Identify the term defined and complete the designated squares.

1	2
3	4
5	6
7	8
9	10

Across

For Scoring

3. The science dealing with randomly occurring events and chance is known as ______. ________
4. A formula from Bayes' theorem solves ______ probability problems. ________
5. His inequality explicitly shows the role of the mean and variance of a distribution in probability limits. ________
8. The ______ of two sets of elements, say A and B, refers to the set of elements in A or in B or in both A and B. ________
9. A method of counting in which X Y is considered to be different from Y X. ________
10. In Chapter 7, the symbol ! means ______. ________

Down

1. A technique of counting in which ordering of the elements is not a consideration. ________
2. An event that is an element of the intersection of two events is called a ______ event. ________
6. His theorem deals with conditional probability. ________
7. An anticipated event in a probability experiment is called a ______. ________

*Any item that has an asterisk before the number indicates it covers optional text material. If such material was not presented in class, you may skip this item. Check with your teacher.

CHAPTER 7

Elementary Probability

Student ______________________________ Score __________

SECTION 3 REHEARSAL EXERCISES

Directions: In the Answers column, write the letter that represents the correct choice.

Answers **For Scoring**

1. The set of all possible outcomes of a probability experiment is called: ______ 1. ______
   (A) conditional probability
   (B) success
   (C) a sample space
   (D) a combination
2. The minimum and maximum proportions that can be used to express probability are: ______ 2. ______
   (A) -1.00; $+1.00$
   (B) 0.00; 1.00
   (C) 0.00; 100.00
   (D) 0; 50
3. Donna, Ricky, Jay, and Leslie are on the high school debate team. What is the probability that either Donna or Jay will be the first debater if the leadoff speaker is chosen by a random draw from among these four? ______ 3. ______
   (A) 0.75
   (B) 0.25
   (C) 0.4
   (D) 0.5
4. Suppose you enter a classroom in which there are seven students, none of whom you know. Three of the students are juniors and four are seniors. You randomly select one of the students. If A represents the event that you select a junior, what is $p(A')$? ______ 4. ______
   (A) $\frac{3}{7}$
   (B) $\frac{1}{7}$
   (C) $\frac{3}{4}$
   (D) $\frac{4}{7}$
5. If the number of elements in $X \cap Y$ is 8 and the number of elements in Y is 40, what is $p(X|Y)$? ______ 5. ______
   (A) 0.20
   (B) 0.08
   (C) 0.05
   (D) 0.50
6. A class contains 10 male and 20 female students. Half the male and half the female students have brown eyes. What is the probability that a student chosen at random is a brown-eyed female? ______ 6. ______
   (A) $\frac{1}{3}$
   (B) $\frac{1}{10}$
   (C) $\frac{1}{2}$
   (D) $\frac{2}{3}$

(continued)

SECTION 3 REHEARSAL EXERCISES (continued)

	Answers	For Scoring
7. Using the scenario of problem 6, what is the probability that the person selected (either male or female) does not have brown eyes; that is, $p(B')$? (A) $\frac{1}{3}$ (B) $\frac{1}{10}$ (C) $\frac{1}{2}$ (D) $\frac{2}{3}$	______	7. ______
8. The number of permutations of 7 objects taken 4 at a time is: (A) 2,520 (B) 840 (C) 765 (D) 35	______	8. ______
9. The number of combinations of letters a, b, c, d, e taken 3 at a time is: (A) 10 (B) 20 (C) 120 (D) 640	______	9. ______
*10. Suppose the sophomore class at Dunbar High was administered a standardized vocational interest inventory. The mean and variance of the distribution of the "social" scale of the inventory were 63 and 121, respectively. Using Tchebycheff's Inequality, determine the maximum probability that a sophomore chosen at random would score above 85 or below 41 on the "social" scale. (A) 0.52 (B) 0.25 (C) 0.18 (D) none of the above	______	10. ______

SECTION 4 PRACTICE PROBLEMS

	Answers	For Scoring
1. A box contains slips of paper on which are printed the letters of the alphabet, one letter per slip. A slip is drawn at random.		
a. What is the probability that the slip will contain the letter J?	______	1a. ______
b. What is the probability that the slip will contain a vowel?	______	1b. ______
c. Given that the slip will contain a vowel, what is the probability that the letter is U; that is, what is $p(U \mid \text{vowel})$?	______	1c. ______
d. If two slips are drawn from the box, what is the probability that the first slip will contain a vowel and the second will contain the letter M?	______	1d. ______

(continued)

Student ______________________ Score __________

SECTION 4 PRACTICE PROBLEMS (continued)

Answers **For Scoring**

2. Consider an experiment in which three different coins are tossed—a quarter, a dime, and a penny.
   a. Complete the list of all possible outcomes, that is, the sample space of the experiment. 2a. ____

Q	D	P
H	H	H
H	H	T
____	____	____
____	____	____
____	____	____
____	____	____
____	____	____
____	____	____

   b. Of the eight outcomes, how many produce three heads? ________ 2b. ____
   c. What is the probability of obtaining three heads? ________ 2c. ____
   d. What is the probability of obtaining exactly one head? ________ 2d. ____
   e. What is the probability of obtaining exactly two heads? ________ 2e. ____
   f. What is the probability of obtaining at least two heads? ________ 2f. ____
   g. What is the probability that all three coins will turn up alike? ________ 2g. ____

3. The junior class of Monterey High, as a fund-raising activity, conducted a candy sale. The boxes of chocolates were the same size and appearance, but the candies had different centers. The distribution of boxes of candy by type of center is as follows:

Nut Center	Fruit Center
Peanut (150 boxes)	Cherry (75 boxes)
Almond (80 boxes)	Raspberry (30 boxes)
Walnut (50 boxes)	Strawberry (30 boxes)

   Suppose a box is to be randomly selected from the inventory to present to the faculty sponsor for her assistance in organizing the money-raising event. Find the following probabilities for the selection.
   a. p(Peanut center) = p(P) = ________ 3a. ____
   b. p(Nut center) = p(P or A or W) = ________ 3b. ____

(continued)

	Answers	For Scoring
c. p(Not Fruit) = $p(F') =$	________	3c. _____
d. p(Raspberry or Walnut) = $p(R \cup W) =$	________	3d. _____
e. p(Almond or Not Nut) = $p(A \cup N') =$	________	3e. _____

4. Denote the event "classified as a senior in high school" by S (underclass or "not senior" event is S′). Twenty-six percent of the student body are seniors. Let B designate the event that a student maintains a B grade average or better, and suppose it is known that 40% of the seniors and 25% of the underclass students maintain this average. Then, the following designations are true:

   ________ 4. _____

   $p(S) = 0.26$
   $p(S') = 0.74$
   $p(B|S) = 0.40$
   $p(B|S') = 0.25$

   If a student has a B or better average, what is the probability that the student is a senior; that is, what is $p(S|B)$?

5. Of the five players on the women's basketball team, your favorite player is Cynthia Chee. You know several things about the team, including the fact that Cynthia is shooting 62% from the free-throw line (she makes 62 out of 100 free throws on the average). You know that the other four players have a combined average of 70% at the free-throw line. If M is the event of making a free throw and C represents Cynthia, then

   ________ 5. _____

   $p(M|C) = 0.62$
   $p(M|C') = 0.70$
   $p(C) = 0.20$
   $p(C') = 0.80$

   Knowing all this, you overhear some students saying that the women's team won the game with a free throw at the end of the game, and you immediately use Bayes' theorem to compute the probability that Cynthia was the one who made the game-winning free throw. What is $p(C|M)$?

6. Assume that 70% of all statistics professors in universities are men. Further assume that 90% of all male statistics professors own a statistical calculator and 80% of all female statistics professors own a statistical calculator. Complete the probability paths in the tree diagram by inserting the probability values for the respective segments. Events are symbolized as:

   Male Professors—M
   Female Professors—M′ (complement of M)
   Owns a Calculator—C
   Does Not Own a Calculator—C′

(continued)

Student ______________________________ Score __________

SECTION 4 PRACTICE PROBLEMS (continued)

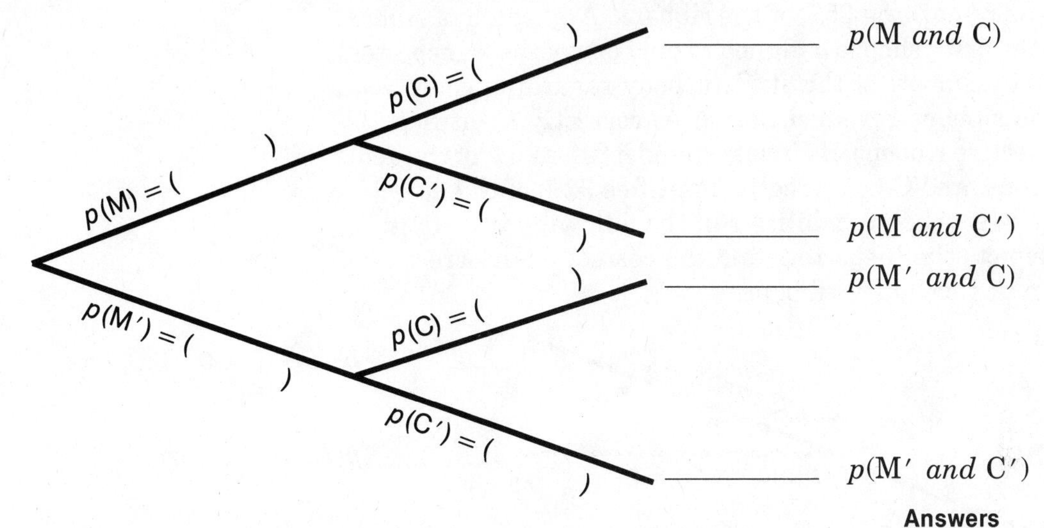

	Answers	For Scoring
a. Sum the probabilities of the four possible outcomes.	______	6a. ____
b. Use the tree diagram to verify the conditional probability: $p(C \mid M) = \frac{p(C \cap M)}{p(M)} =$	______	6b. ____
c. What is the probability that a statistics professor selected at random will own a calculator? [Hint: $p(M \cap C) + p(M' \cap C)$.]	______	6c. ____

7. A survey of the school parking lot revealed that 40% of the cars had faculty parking stickers and the rest had student parking stickers. Further, 5% of the faculty cars had tape decks and 30% of the student cars had tape decks. Complete the tree diagram and determine the probability that a randomly selected car from the lot would have a tape deck.

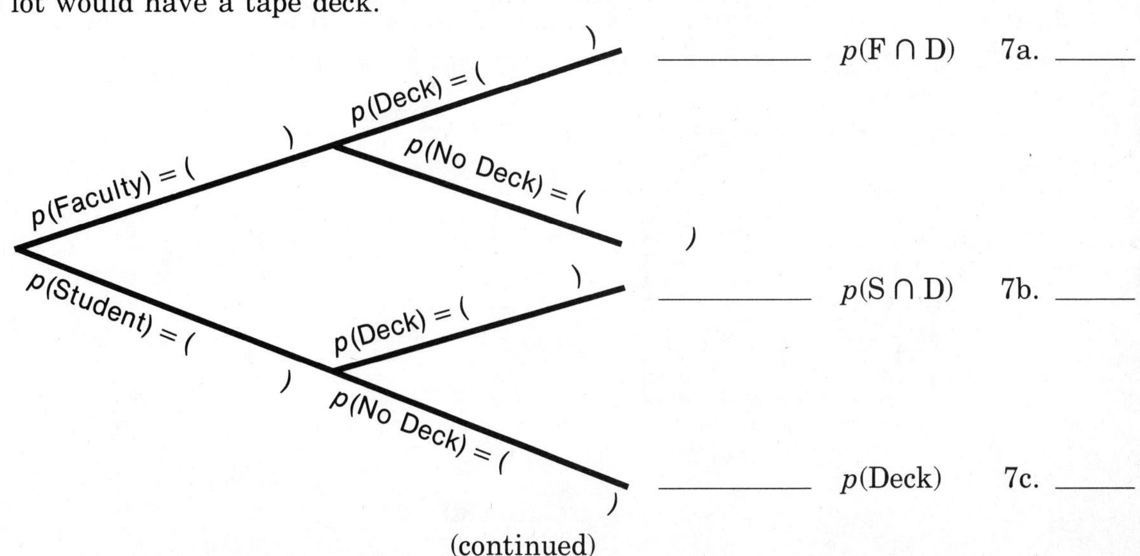

(continued)

 Answers **For Scoring**

8. The senior class at Highland High School is responsible for designing, printing, and marketing school spirit ribbons prior to each varsity athletic event. Three class members have been chosen to serve as quality control inspectors of the ribbons. A.A. inspects 35% of the ribbons, B.B. analyzes 40%, and C.C. checks the remaining 25% of the ribbons. Any batch of ribbons leaving the print shop is assigned to only one of these inspectors. Results of a survey of the student body regarding complaints about the ribbons revealed that A.A. correctly identifies 95% of the defective ribbons, B.B. correctly identifies 90% of the defective ribbons, and C.C. correctly identifies 92% of the defective ribbons. Insert the probabilities in the following tree diagram, which depicts the inspectors and the correct or incorrect identification of the faulty ribbons.

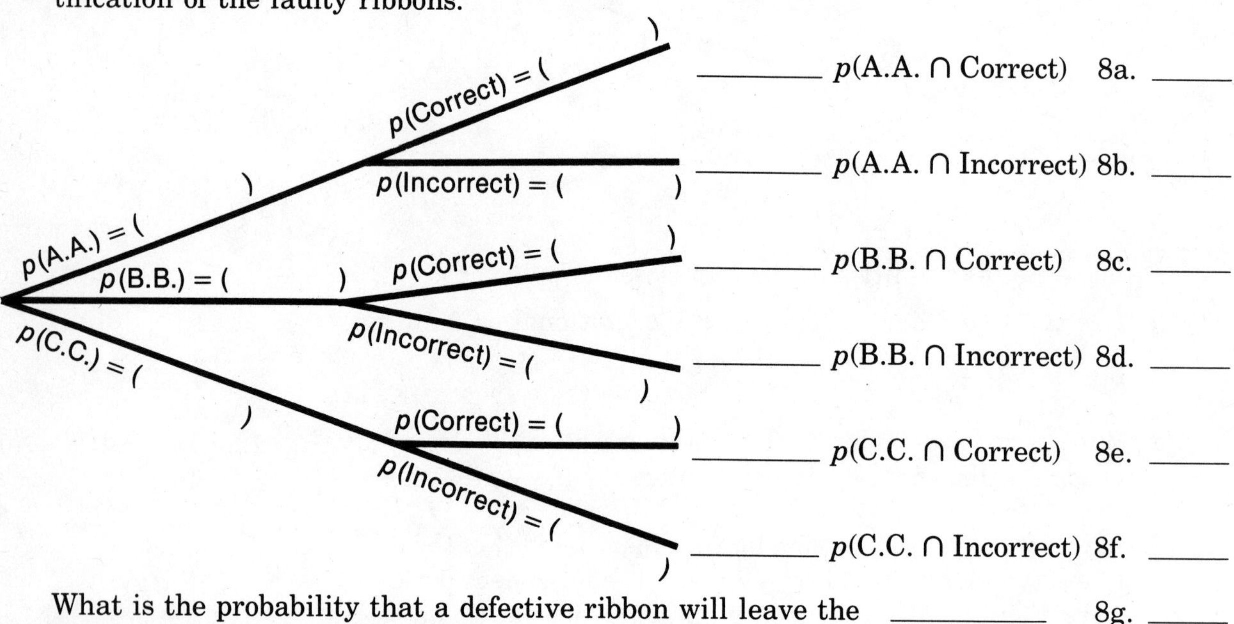

_______ $p(\text{A.A.} \cap \text{Correct})$ 8a. _____

_______ $p(\text{A.A.} \cap \text{Incorrect})$ 8b. _____

_______ $p(\text{B.B.} \cap \text{Correct})$ 8c. _____

_______ $p(\text{B.B.} \cap \text{Incorrect})$ 8d. _____

_______ $p(\text{C.C.} \cap \text{Correct})$ 8e. _____

_______ $p(\text{C.C.} \cap \text{Incorrect})$ 8f. _____

What is the probability that a defective ribbon will leave the print shop without being detected? _______ 8g. _____

9. Given a sample space (Universe) U and an event u defined by the following diagrams (a. through d.), find the probability that a randomly chosen point in U belongs to u; that is, $p(u) = ?$

a. $p(u) = \dfrac{\text{area of } u}{\text{area of } U} =$ _______ 9a. _____

u
U
10 cm
10 cm

(continued)

Student ____ Score __________

SECTION 4 PRACTICE PROBLEMS (continued)

Answers | **For Scoring**

b.

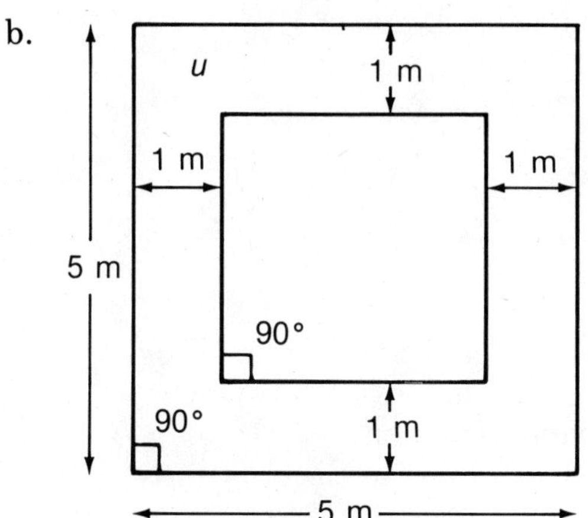

$p(u) = \dfrac{\text{area of } u}{\text{area of } U} =$ __________ 9b. _____

c. $p(u) = \dfrac{\text{length of } u}{\text{length of } U} =$ __________ 9c. _____

d. $p(u) = \dfrac{\text{area of } u}{\text{area of } U} =$ __________ 9d. _____

U: (circular area)

u: (shaded area)

10. A high-voltage electrical line that spans a distance of 65 miles from the remote generating plant to the nearest community is damaged by a storm at a random point. A crew from the generating plant will repair the damage if it is within 30 miles of the generator; otherwise a crew from the community will be dispatched.

    a. What is U (the sample space)? __________ 10a. _____

    b. What is u? __________ 10b. _____

    c. Find the probability that the damage was within 30 miles of the generating plant. __________ 10c. _____

(continued)

Answers **For Scoring**

11. A field is in the shape of a right triangle and has sides of 0.4 km and 0.5 km as shown. ________ 11. _____

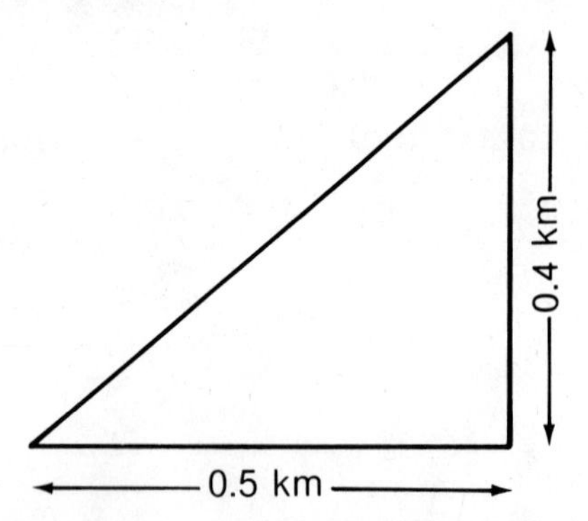

A security wire has been installed around the perimeter of the field. A small animal trips the security wire at a random location. What is the probability that the security violation is on the hypotenuse of the perimeter?

CHAPTER 8
Probability Distributions

Student ______________________ Total Score ________

SECTION 1 CONCEPT COMPREHENSION

Directions: Complete the following by writing the correct word or words in the respective blanks to the right.

Many probability experiments involve a sequence of independent trials in which each trial has only two outcomes. These are referred to as __(1)__ or Bernoulli experiments. In the study of probability, these outcomes are generally classified as __(2)__ or failures. The number of successes in a Bernoulli experiment is a __(3)__ variable, because it is assumed that the outcome of any particular trial is a random event.

The __(4)__ distribution is the probability distribution of s for Bernoulli experiments. This distribution is defined by n, the number of trials, and by p, the probability of a __(5)__ in a single trial. Given p and n for the distribution, the mean (μ) and the variance (σ) can be computed by __(6)__ and __(7)__ (where $q = 1 - p$), respectively. Coefficients of an expanded binomial in which $p = q = \frac{1}{2}$ can be arranged in a configuration known as __(8)__ triangle for convenient reference.

In a binomial experiment in which the number of trials (n) is large and p is close to $\frac{1}{2}$, the __(9)__ distribution becomes a very good approximation to the binomial. The area under the curve of this continuous distribution, when expressed as a proportion, is interpreted as relative frequency or as __(10)__.

If events are occurring at a particular rate, on the average, the __*(11)__ probability distribution can be used to find the probability that the event will occur a certain number of times within a particular time span. In this distribution the value of the mean is the same as the value of the __*(12)__ of the distribution. Further, if p is small and n large, this distribution is a reasonably good approximation to the binomial, and in such cases __*(13)__ $= np$.

	Answers	For Scoring
1.	________	1. ____
2.	________	2. ____
3.	________	3. ____
4.	________	4. ____
5.	________	5. ____
6.	________	6. ____
7.	________	7. ____
8.	________	8. ____
9.	________	9. ____
10.	________	10. ____
11.	________	11. ____
12.	________	12. ____
13.	________	13. ____

*Any item that has an asterisk before the number indicates it covers optional text material. If such material was not presented in class, you may skip this item. Check with your teacher.

SECTION 2 VOCABULARY PRACTICE

Directions: Identify the term defined and complete the designated squares.

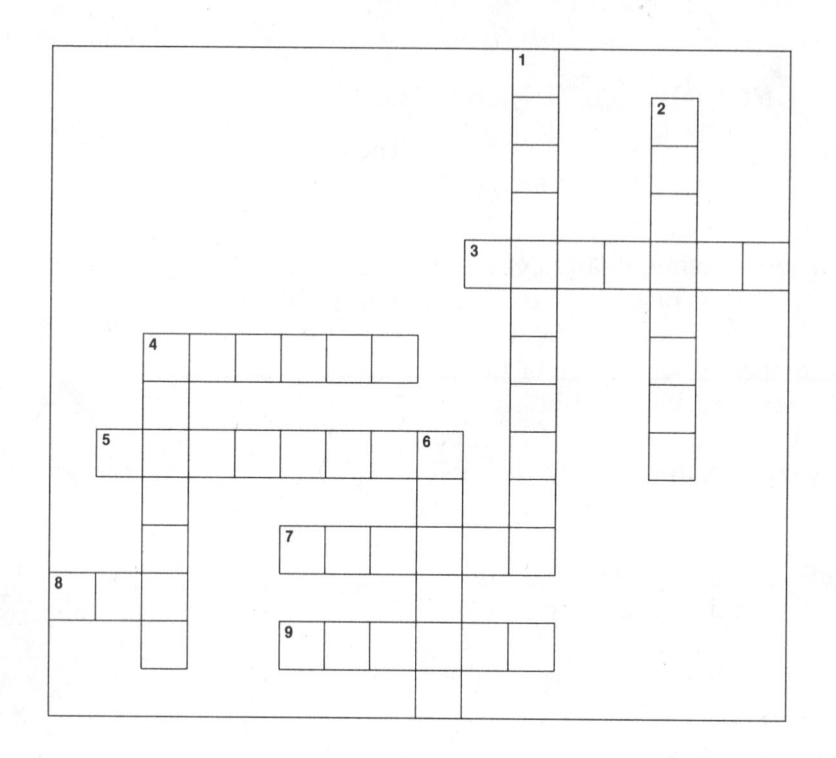

Across | **For Scoring**

3. A frequency distribution curve of a random variable. ________
4. His triangle may be used to expand a binomial expression when $p = \frac{1}{2}$. ________
5. A Bernoulli experiment is a ______ experiment. ________
7. A bell-shaped curve. ________
8. In a binomial experiment, only this many outcomes are possible. ________
9. The number of possible "successes" of an experiment is called a ______ variable. ________

Down

1. The normal curve is ______ about the mean; that is, it is distributed the same above and below the mean. ________
2. A random ______ is sometimes called a variate. ________

*4. This distribution has a mean equal to its variance. ________

*6. In a Poisson distribution, the Greek letter ______ (λ) is the mean. ________

Probability Distributions

Student ______________________________ Score __________

SECTION 3 REHEARSAL EXERCISES

Directions: In the Answers column, write the letter that represents the correct choice.

	Answers	For Scoring
1. If X is a random variable indicating the number of heads showing after two coins are tossed (that is, 0, 1, or 2), what are the probabilities associated with each of the X's? In other words, what are $p(X = 0)$, $p(X = 1)$ and $p(X = 2)$, respectively? (A) 1, 2, 1 (B) 0.25, 0.25, 0.25 (C) 0.25, 0.50, 0.75 (D) 0.25, 0.50, 0.25	______	1. ______
2. Suppose each of five students has a 50% chance of being selected for a summer institute in drama. Using the binomial formula, what is the probability that at least three of these students will be selected? (A) $\frac{1}{2}$ (B) $\frac{10}{16}$ (C) $\frac{5}{32}$ (D) $\frac{25}{32}$	______	2. ______
3. A fair die is rolled seven times. A toss is considered a success if either a 5 or a 6 appears. For a binomial experiment, what are the respective values of p and q? (A) $\frac{1}{2}, \frac{1}{2}$ (B) $\frac{1}{6}, \frac{1}{6}$ (C) $\frac{1}{3}, \frac{2}{3}$ (D) $\frac{1}{6}, \frac{5}{6}$	______	3. ______
4. If a fair die is rolled seven times as in the previous problem and the appearance of either a 5 or a 6 is a success (s), what is $p(s = 3)$? (A) $\frac{688}{2,187}$ (B) $\frac{560}{2,187}$ (C) $\frac{2,059}{2,187}$ (D) $\frac{128}{2,187}$	______	4. ______
5. The mean of a binomial distribution when $n = 64$ and $p = \frac{1}{2}$ is: (A) 16 (B) 32 (C) 4 (D) 64	______	5. ______

(continued)

Answers **For Scoring**

6. Suppose the Friendship High School chess team has won 75% of its matches against district competition during the fall semester. There are 36 games scheduled for the second round of competition during the spring term. Assuming that the *mean* number of successes over many such semesters would be the best estimate of the number of wins for Friendship for the spring semester, what is the estimated number of wins? ______ 6. ______
(A) 18
(B) 24
(C) 27
(D) 32

7. Suppose final exam scores in physics were normally distributed. If one student from physics is selected at random, what is the approximate probability that the student scored within one standard deviation of the class mean? ______ 7. ______
(A) 0.34
(B) 0.50
(C) 0.16
(D) 0.68

*8. The student body president has, on the average, five requests per month to make a speech to a school organization or to a community civic organization. Using the Poisson distribution, what is the probability that she will get exactly two requests during any given month? ______ 8. ______
(A) 0.0842
(B) 0.2022
(C) 0.1179
(D) 0.0067

*9. The average number of accidental drownings per year in the United States is approximately 3 per 100,000 population. Using the Poisson distribution, what is the probability that in a city of 200,000 population, there will be fewer than 3 accidental drownings per year? ______ 9. ______
(A) 0.0620
(B) 0.1240
(C) 0.0149
(D) 0.0446

10. The person credited with an easily computed triangular array of binomial coefficients when $p = q = \frac{1}{2}$ is: ______ 10. ______
*(A) Poisson
(B) Bernoulli
(C) Pascal
(D) Bayes

Probability Distributions

Student ______________________________ Score __________

SECTION 4 PRACTICE PROBLEMS

	Answers	For Scoring
1. Find the probability that a normally distributed standard z-score lies:		
a. below -1.96	________	1a. _____
b. above -1.00	________	1b. _____
c. between 1.5 and 2.6	________	1c. _____
d. between -1.2 and $+3.0$	________	1d. _____
2. The probability that Hector, the leading marksman on the Meadow High School trap-shooting team, will hit a certain target is $\frac{2}{3}$. If he takes nine shots, find the probability that Hector will hit the target seven or more times (use the binomial distribution).	________	2. _____
3. In a recent survey, 23% of the high school seniors in the U.S. indicated they would like to become teachers. If a sample of 50 students is randomly taken from a senior class, determine the probability that:		
a. 10 or fewer would like to become teachers	________	3a. _____
b. more than 10 but fewer than 20 would like to become teachers	________	3b. _____
*4. Students in the woodshop class can turn out an average of four standard bookshelves per day. Use the Poisson distribution to determine the probability that on a particular day five or more bookshelves will be built by the class. [Hint: Determine $p(s \leq 5)$, then subtract from 1.0.]	________	4. _____
*5. If on the average 2% of the students in a school are left-handed, find the probability that fewer than 3 of the 100 junior class members are left-handed.	________	5. _____

(continued)

6. Consider a binomial experiment for which $p = q = \frac{1}{2}$ and $n = 8$.
   a. Compute μ and σ. ______ 6a. ______

   b. Find the probability of seven or more successes or one or fewer successes; that is, $p(s \geq 7) + p(s \leq 1)$. In other words, find the probability that the number of successes will depart from the mean by 3 or more. ______ 6b. ______

   *c. In b. you found the probability that the value of the random variable s would depart from μ by at least 3 units; that is, $\epsilon = 3$. Apply Tchebycheff's Inequality to the problem and verify that the probability in b. is $\leq \frac{\sigma^2}{\epsilon^2}$. ______ 6c. ______

7. Suppose attendance at a high school is normally distributed with a mean of 2,450 and a standard deviation of 36.
   a. Assuming other factors are equal, what is the probability that on a randomly selected school day the attendance will be more than 2,500? ______ 7a. ______

   b. What is the probability that on a randomly selected school day the attendance will be greater than 2,425? ______ 7b. ______

   c. What is the probability that on a randomly selected day the attendance will be between 2,400 and 2,500? ______ 7c. ______

8. A single die will be rolled 10 times. What is the probability of getting exactly two aces (ones)? ______ 8. ______

9. A 50-item multiple choice exam has four possible responses per item. If a student knows nothing about the content, what is the probability of getting a score of 30 or higher (30 or more correct)? ______ 9. ______

10. The Mesa High School debate team has a $\frac{4}{7}$ probability of winning whenever it debates an opponent. Suppose the team schedules four debates during the month of January. What is the probability that it will win:
    a. exactly two debates? ______ 10a. ______

    b. more than two debates? ______ 10b. ______

    c. all four debates? ______ 10c. ______

    d. none of the four debates? ______ 10d. ______

    e. What are the odds of Mesa winning a debate? ______ 10e. ______

    ...hat are the odds *against* Mesa winning a debate? ______ 10f. ______

(continued)

Student ______________________________ Score __________

SECTION 4 PRACTICE PROBLEMS (continued)

	Answers	For Scoring
11. The Mesa debate team consists of six students. Assume that the probability of any particular member being female is $\frac{1}{2}$. What is the probability that:		
a. there are three male and three female student debaters?	______	11a. ____
b. there are fewer boys than girls?	______	11b. ____
12. Complete the next five rows of Pascal's triangle.		12. ____

1

1 1

1 2 1

1 3 3 1

___ ___ ___ ___ ___

___ ___ ___ ___ ___ ___

___ ___ ___ ___ ___ ___ ___

___ ___ ___ ___ ___ ___ ___ ___

___ ___ ___ ___ ___ ___ ___ ___ ___

Applied Sampling

Student ____ Total Score __________

SECTION 1 CONCEPT COMPREHENSION

Directions: Complete the following by writing the correct word or words in the respective blanks to the right.

The superintendent of Jefferson Public Schools is interested in the amount of inservice education the 420 teachers in the district have received in the past two years. The superintendent decides to survey only a representative portion of all of the teachers. Statistically speaking, all of the teachers in the district represent a __(1)__, while the subset to be surveyed is known as a __(2)__. For valid inferences from the survey results on the smaller group to be made to all teachers, the subset of teachers must be __(3)__ of the entire group of teachers in the district. Selecting the subset to be surveyed from the teachers who attend an open-house reception for the new high-school principal would probably yield a __(4)__ subset, because elementary-school teachers would not be in attendance and thus would not have a chance to respond to the superintendent's survey.

To select the sample, the superintendent will use a scheme that gives each of the 420 teachers in the district an equal chance of being selected. This is called a __(5)__ sampling technique. With such a method, the number of teachers in the sample would need to be at least __(6)__ for the superintendent to be 95% confident that extreme cases will be included from the district's 420 teachers.

The superintendent will determine the mean number of hours of inservice training over the past two-year period as reported by the teachers participating in the survey. This average, based on the sample data, is called a __(7)__ and will be used to estimate the mean for all teachers, which is called a __(8)__. This estimation or generalization process is known as statistical __(9)__ and is employed extensively in our society every day.

	Answers	For Scoring
1.	______________	1. ____
2.	______________	2. ____
3.	______________	3. ____
4.	______________	4. ____
5.	______________	5. ____
6.	______________	6. ____
7.	______________	7. ____
8.	______________	8. ____
9.	______________	9. ____

SECTION 2 VOCABULARY PRACTICE

Directions: Identify the term defined and complete the designated squares.

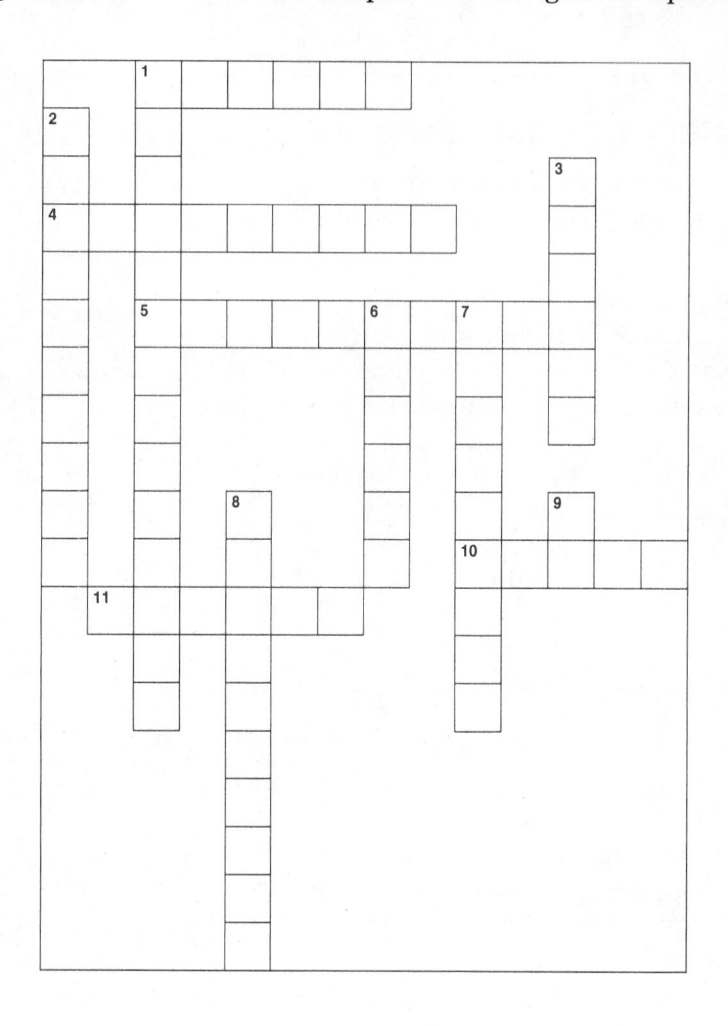

Across

For Scoring

1. A sample in which every member of the population has an equal opportunity to be included. ________
4. A characteristic of a population and usually unknown. ________
5. ______ are results of analyses on sample data. ________
10. Random samples provide an ______ opportunity for each element of the population to be selected. ________
11. Samples that are not representative of a population. ________

Down

1. Samples that reflect the same properties as the population they represent. ________
2. A "universal" set. ________
3. A portion of a set of elements. ________
6. Part of a population. ________
7. Generalization from a sample to a population. ________
8. Not random. ________
9. μ ________

Applied Sampling

Student ______________________________ Score ________

SECTION 3 REHEARSAL EXERCISES

Directions: In the Answers column, write the letter that represents the correct choice.

Answers | For Scoring

1. A measure taken on a sample is called a: ______ 1. _____
   (A) parameter
   (B) discrete variable
   (C) statistic
   (D) nominal variable
2. The procedure for using data from a sample to answer questions about a population is called: ______ 2. _____
   (A) descriptive statistics
   (B) inferential statistics
   (C) probability
   (D) measurement
3. Analogy: ______ is to "population" as "statistics" is to ______. ______ 3. _____
   (A) parameter; sample
   (B) sample; parameter
   (C) sample; random
   (D) parameter; inference
4. Creating a sample of students by starting with the second name in the student directory and selecting every fifteenth name best describes what kind of sample? ______ 4. _____
   (A) random
   (B) cluster
   (C) stratified random
   (D) systematic
5. μ symbolizes a ______, whereas $\overline{X}$ represents a ______. ______ 5. _____
   (A) sample; statistic
   (B) parameter; statistic
   (C) sample; population
   (D) population standard deviation; population mean
6. Samples of "convenience" are probably: ______ 6. _____
   (A) unbiased
   (B) representative
   (C) reliable
   (D) biased
7. If the governors of the 50 states were defined as a population of interest, their mean age would be a(n): ______ 7. _____
   (A) parameter
   (B) statistic
   (C) estimate of the parameter
   (D) inferred quantity

(continued)

SECTION 3 REHEARSAL EXERCISES (continued)

	Answers	For Scoring
8. A major TV network claimed that 45% of the television viewing audience watched a certain "special" presentation. The 45% figure is: (A) a fact (B) a statistic (C) a sample (D) misleading	______	8. _____
9. For a researcher to be 95% confident that extremes are represented, the minimum number of subjects in a sample randomly drawn from a population of 50,000 people would be: (A) 50 (B) 381 (C) 191 (D) 3,507	______	9. _____
10. Generally, as the size of the population gets larger, the sample size necessary to provide representativeness: (A) gets smaller (B) gets larger at the same rate as the increase of the population size (C) gets larger, but at a rate less than the rate of increase in the population size (D) remains constant at about 10% of the size of the population	______	10. _____

SECTION 4 PRACTICE PROBLEMS

	Answers	For Scoring
1. If N = the size of a population and s = the recommended sample size, then $$s = \frac{0.96025(N)}{0.0025(N-1) + 0.96025}$$ (rounded to the nearest whole number). Assuming samples are selected at random, compute the sample sizes necessary for populations with sizes of:		
a. $N = 5{,}500$	______________	1a. _____
b. $N = 15{,}000$	______________	1b. _____
c. $N = 200{,}000$	______________	1c. _____
2. Eight people are randomly selected from each city block to form a sample of a city's population. This sampling procedure is called ______.	______________	2. _____

(continued)

Student ____________________ Score ________

SECTION 4 PRACTICE PROBLEMS (continued)

Answers | For Scoring

3. In a certain high school, suppose there are 250 seniors, 200 juniors, and 240 sophomores. If you decide to use a stratified (by grade level) random sampling technique to obtain a sample that you can be 95% confident has included extreme cases from *each of the three grade levels*, what will be the minimum total sample size you will need from the high school? ________ 3. ____

4. With reference to problem 3, suppose the extremes in each grade level are not an issue, but a sample representative of the entire student body is desired. What is the minimum size sample needed? ________ 4. ____

5. If a sample of insufficient size is selected from a population and if a measure of variability is calculated on the variable of interest, is the variability *statistic* apt to be a valid estimate of the *parameter*? Why or why not? 5. ____

____

____

____

____

____

____

____

6. Why is sample size alone not a sufficient condition for securing a representative sample? 6. ____

____

____

____

____

____

____

____

(continued)

For Scoring

7. Suppose that swine flu vaccinations were provided for the elderly and infirm (a high-risk group for flu). Suppose that this group who had taken the vaccinations turned out to have a higher incidence of flu than the general population. Would this provide evidence that the vaccine was ineffective or perhaps even caused increased rates of flu? Why or why not?

7. ______

8. Are the phrases "representative sample" and "random sample" synonymous? Explain your answer.

8. ______

9. Distinguish between a parameter and a statistic.

9. ______

10. Inserted in a certain magazine will be a self-addressed, postage-paid postcard with 10 survey questions. The circulation of the magazine is 75,000 in a particular state and the sponsors, who want the opinions of residents of the state, have used a full page to describe the purpose and procedure of the survey. Readers are asked to complete the survey and mail the postcard back to the sponsor. Decide what the intended population was and evaluate the validity of the sample from a statistical perspective.

10. ______

(continued)

Student ______________________ Score ________

SECTION 4 PRACTICE PROBLEMS (continued)

For Scoring

The principal at Wilson High School wants to know the average IQ of her student body. However, individual testing requires about $1\frac{1}{2}$ hours of student time per test and another hour of examiner time for scoring and interpreting the test. Therefore the procedure is too expensive and time consuming for the entire student body to be tested. Hence the principal will test a sample of students. For problems 11 through 15, comment on the likelihood that the technique will provide a representative sample of students for the testing project.

11. Select every tenth student passing through the cafeteria line at noon. 11. _____

12. Test the first 50 students arriving at school on a given day. 12. _____

13. Select every third student of those volunteering during the sixth class period of the day. 13. _____

14. Randomly select 365 students from a listing of all students. 14. _____

(continued)

For Scoring

15. Test all students attending art class on a given day.

15. _____

____

____

____

____

____

CHAPTER 10
Estimation

Student ______________________________ Total Score ____________

SECTION 1 CONCEPT COMPREHENSION

Directions: Complete the following by writing the correct word or words in the respective blanks to the right.

The process of generalizing results of sample data to populations is known as ___(1)___. The results of computations on the samples are ___(2)___ and are used to estimate population ___(3)___. For example, the sample ___(4)___ is used to estimate μ. When a single value is used as a statistic to estimate a population characteristic, it is called a(n) ___(5)___ estimate. In the current chapter you learned that s (computed with $N - 1$) is a(n) ___(6)___ estimate of σ. However, because we know that some sampling ___(7)___ will be involved, a range of values called a(n) ___(8)___ interval is used for estimation.

To increase the level of confidence, the width of the confidence interval must be ___(9)___. Alternatively, you can achieve the same level of confidence with a decreased interval width if ___(10)___ is increased. The use of confidence intervals is made possible by the ___(11)___ theorem. This theorem states that the standard deviation of a sampling distribution of the mean, called a standard ___(12)___ of the mean, can be estimated by dividing the standard deviation of the original distribution by the square root of ___(13)___. Further, the form of the sampling distribution is a(n) ___(14)___ curve, which permits the use of tabled values of the area under the curve. Although the level of confidence chosen is arbitrary, the 95% and 99% confidence intervals are very popular, and the computations of these intervals make use of the normal curve z-scores of ___(15)___ and ___(16)___ for the respective intervals.

	Answers	For Scoring
1.	____________	1. ____
2.	____________	2. ____
3.	____________	3. ____
4.	____________	4. ____
5.	____________	5. ____
6.	____________	6. ____
7.	____________	7. ____
8.	____________	8. ____
9.	____________	9. ____
10.	____________	10. ____
11.	____________	11. ____
12.	____________	12. ____
13.	____________	13. ____
14.	____________	14. ____
15.	____________	15. ____
16.	____________	16. ____

SECTION 2 VOCABULARY PRACTICE

Directions: Identify the term defined and complete the designated squares.

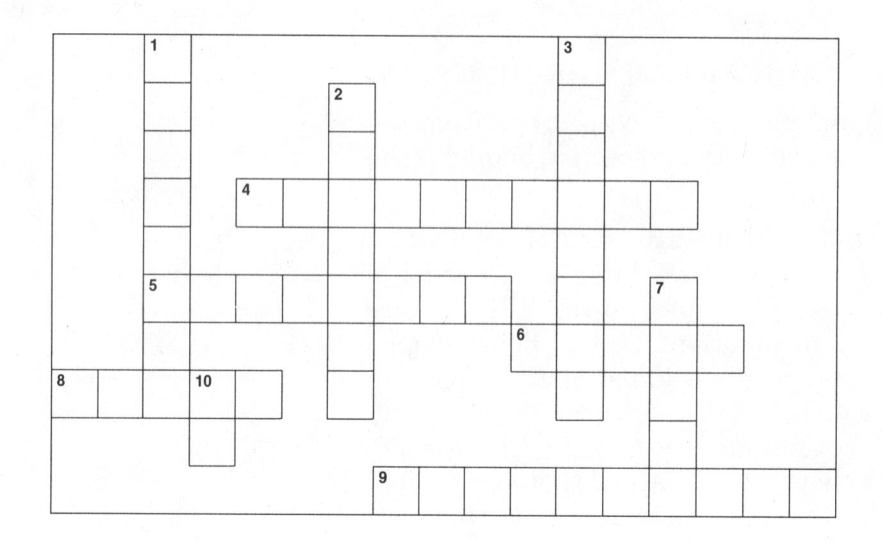

Across | **For Scoring**

4. ______ intervals are used to set upper and lower bounds on estimates of a population mean. ________
5. An ______ estimate takes sampling error into account. ________
6. Sampling ______ is quantified by the standard error of the mean. ________
8. The Greek letter (σ) used to symbolize the population standard deviation. ________
9. The primary topic of Chapter 10 is ______. ________

Down

1. A theoretical distribution of sample statistics is known as a ______ distribution. ________
2. The ______ limit theorem provides information about the standard deviation of a sampling distribution of means. ________
3. The ______ error of the mean is a standard deviation of a sampling distribution of means. ________
7. A statistical inference using a single value as the best estimate is a ______ estimate. ________
10. The mean of all possible random sample means. ________

Student ____ Score ____________

SECTION 3 REHEARSAL EXERCISES

Directions: In the Answers column, write the letter that represents the correct choice.

Answers **For Scoring**

1. When one computes a confidence interval for the mean, the upper and lower limits of the interval represent an interval estimate of the: ______ 1. ______
   (A) sample mean
   (B) sample standard deviation
   (C) population mean
   (D) population standard deviation

2. For a particular variable X, which of the following quantities will represent the smallest (shortest) interval on the baseline? ______ 2. ______
   (A) standard deviation of X
   (B) range of X
   (C) standard error of the mean of X
   (D) $\mu \pm 1\sigma$

3. The standard deviation of a distribution of all possible sample means from a population is called the: ______ 3. ______
   (A) standard error of the mean
   (B) 95% confidence interval
   (C) population mean
   (D) none of the above

4. The 99% confidence interval for the mean is ______ the 95% confidence interval for the same mean. ______ 4. ______
   (A) larger than
   (B) smaller than
   (C) equal to
   (D) less than or equal to

5. A sample of 10 scores has $\overline{X} = 40$ and $\Sigma(X - \overline{X})^2 = 90$. The best estimate of σ^2 from this information is: ______ 5. ______
   (A) 50
   (B) 5
   (C) 9
   (D) 10

6. Assume that you are going to draw a random sample of 64 cases from a population with a mean of 25 and a standard deviation of 10. What is the probability of obtaining a sample with a mean $(\overline{X})$ between 24 and 26 for the random sample? ______ 6. ______
   (A) $\frac{25}{256}$
   (B) 0.20
   (C) 0.58
   (D) 0.79

(continued)

SECTION 3 REHEARSAL EXERCISES (continued)

	Answers	For Scoring
7. Suppose that a randomly selected group ($N = 81$) of 11-year-old students from an urban school system is asked to cancel (cross out) as many vowels in a printed paragraph as possible within a three-minute span of time. The sample mean is 36 vowels, and the standard deviation is 9. The 95% confidence interval for the mean is: (A) {8.0 to 10.0} (B) {34.04 to 37.96} (C) {30.0 to 37.0} (D) {36.04 to 37.12}	______	7. _____
8. If a sample of 25 scores is drawn from a population with a mean of 55 and a standard deviation of 15, approximately 84% of the means of samples of size 25 would lie below: (A) 52 (B) 55 (C) 58 (D) 53	______	8. _____
9. If a 95% confidence interval = {86.52 to 89.48}, which of the following could be a 99% confidence interval for the same data? (A) {86.98 to 89.02} (B) {86.37 to 89.63} (C) both A and B (D) neither A nor B	______	9. _____
10. A theoretical distribution consisting of the mean scores of all possible random samples of a given size is a: (A) highly skewed distribution (B) distribution of all possible standard deviations (C) theoretical distribution of the samples (D) sampling distribution of the mean	______	10. _____

SECTION 4 PRACTICE PROBLEMS

Following are 10 scores on a statistics exam with 20 possible points. Perform the indicated computations.

Score	X^2
17	________
15	________
16	________
18	________
20	________
16	________
16	________
15	________
16	________
12	________

(continued)

Student ______________________ Score ________

SECTION 4 PRACTICE PROBLEMS (continued)

	Answers	For Scoring
1. $\Sigma X =$ ______	______	1. ______
2. $\Sigma X^2 =$ ______	______	2. ______
3. $(\Sigma X)^2 =$ ______	______	3. ______
4. $\Sigma x^2 = \Sigma X^2 - \frac{(\Sigma X)^2}{N} =$ ______	______	4. ______
5. $s = \sqrt{\frac{\Sigma x^2}{N-1}} =$ ______	______	5. ______
6. $s_{\overline{X}} = \frac{s}{\sqrt{N}} =$ ______	______	6. ______
7. 95% confidence interval = ______	______	7. ______

Given the following set of numbers, perform the indicated calculations.

Y
11
20
15
15
13
17
15
15
18
19

8. $\Sigma y^2 = \Sigma Y^2 - \frac{(\Sigma Y)^2}{N} =$ ______	______	8. ______
9. $s = \sqrt{\frac{\Sigma y^2}{N-1}} =$ ______	______	9. ______
10. $s_{\overline{Y}} = \frac{s}{\sqrt{N}} =$ ______	______	10. ______
11. 95% C.I. = ______	______	11. ______

(continued)

	Answers	For Scoring
12. 90% C.I. = ______	______	12. ______
13. 80% C.I. = ______	______	13. ______
14. 68% C.I. = ______	______	14. ______
15. 99% C.I. = ______	______	15. ______

16. Explain the meaning of the 99% confidence interval calculated in problem 15. 16. ______

______

______

______

______

______

17. Assume that the average height of five varsity cheerleaders at Clear Spring High School is 165 centimeters (cm). One more cheerleader has been elected, and the new average height is 167 centimeters for the six cheerleaders.

   a. How tall is the newly elected cheerleader? ______ 17a. ______

   b. What concept is illustrated by the solution to a.? 17b. ______

   ______

   ______

   ______

   ______

   ______

18. In a random sample of 47 students from the junior class of Rockview High School, it was found that 14 had visited the Grand Canyon.

   a. Given these statistics, what is the point estimate of the percentage of juniors at Rockview who have visited the Grand Canyon? ______ 18a. ______

   b. If there are 163 juniors at Rockview, about how many of them have been to the Grand Canyon? ______ 18b. ______

Hypothesis Testing

Student ______________________________ Total Score __________

SECTION 1 CONCEPT COMPREHENSION

Directions: Complete the following by writing the correct word or words in the respective blanks to the right.

The standard error of the mean is necessary to estimate a confidence interval for a population __(1)__ or to test a(n) __(2)__. In turn, an estimate of the population __(3)__ and the sample size are necessary for computing $s_{\overline{X}}$. Hypothesis testing does not prove a conjecture or hypothesis, but statistics do tell the __(4)__ of committing a type I error when statistical significance is achieved. A type II error can occur only when we __(5)__ to reject the null hypothesis (H_0). Prior to rejecting a null hypothesis, a statistician must be assured that values as deviant from the population mean as the obtained sample mean have a very __(6)__ probability of occurring if the null hypothesis is true.

A(n) __(7)__-test can be used as a statistical test of the difference between two means. The calculated value of z is compared to a tabled or __(8)__ value of z to determine if the z-score falls in an area of __(9)__. The __(10)__ level, symbolized α, determines the area of rejection and specifies the probability of committing a __(11)__ when the null hypothesis is rejected.

Rejection of the null hypothesis provides confirming support for the __(12)__ hypothesis. Failure to reject the null hypothesis does not imply that the alternate hypothesis should be __(13)__. In the current chapter, the major focus was on testing for a difference between a μ and a __(14)__.

Answers	For Scoring
1. ________	1. ____
2. ________	2. ____
3. ________	3. ____
4. ________	4. ____
5. ________	5. ____
6. ________	6. ____
7. ________	7. ____
8. ________	8. ____
9. ________	9. ____
10. ________	10. ____
11. ________	11. ____
12. ________	12. ____
13. ________	13. ____
14. ________	14. ____

SECTION 2 VOCABULARY PRACTICE

Directions: Identify the term defined and complete the designated squares.

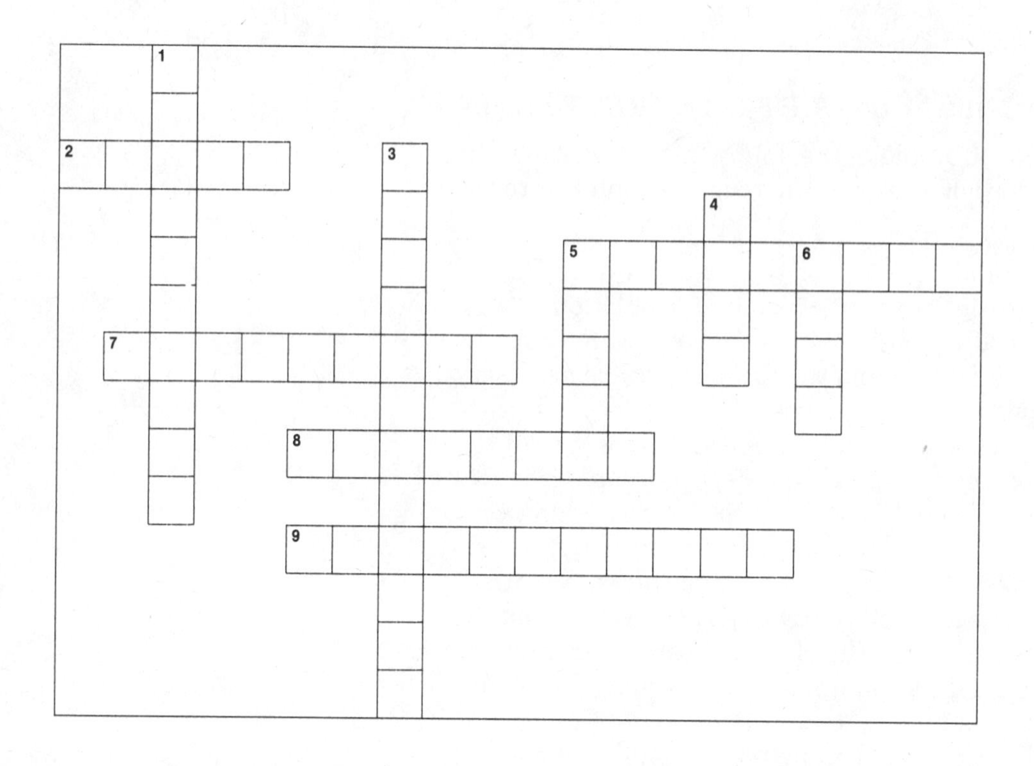

Across

For Scoring

2. An error created by rejecting a null hypothesis that should not have been rejected. ________
5. Rejection of the null hypothesis provides support for the ______ hypothesis. ________
7. The critical value of the z-test designates the area of ______ on the normal curve. ________
8. The tabled value of z that separates the normal curve into α and $1 - \alpha$ proportions. ________
9. ______ significance is achieved when the absolute value of the calculated value of z exceeds the absolute value of the critical (tabled) value of z. ________

Down

1. A conjectural statement about the relationship between variables. ________
3. The 0.05 level of ______. ________
4. The probability of committing a type II error when the null hypothesis has not been rejected. ________
5. The predetermined significance level. ________
6. The ______ hypothesis is either rejected or not rejected by the statistical test. ________

Hypothesis Testing

Student ______________________________ Score ____________

SECTION 3 REHEARSAL EXERCISES

Directions: In the Answers column, write the letter that represents the correct choice.

	Answers	For Scoring
1. Failing to reject a null hypothesis that is false results in: (A) no error (B) a type I error (C) a type II error (D) both a type I and a type II error	______	1. _____
2. If a null hypothesis is rejected at the 0.05 level of significance, then: (A) there is a 5% risk of committing a type I error (B) there is a 95% risk of committing a type I error (C) the observed phenomenon is probably just a random occurrence (D) there exists a 5% probability that a type II error has been committed	______	2. _____
3. A type I error occurs when an experimenter erroneously ______ a null hypothesis that she or he is testing. (A) accepts (B) rejects (C) states (D) fails to reject	______	3. _____
4. Johnny answered all eight items on a true-false exam correctly by simply flipping a coin and counting a head as True and a tail as False. Because it was a well-known fact that Johnny had not studied the material and because the probability of getting such results is about 0.004 ($p < 0.004$) by chance, the teacher concluded that Johnny had cheated on the test. In statistical terminology, the teacher had: (A) committed a type I error (B) committed a type II error (C) inappropriately used statistical methods (D) committed no error	______	4. _____
5. Given the following data on height measurement for a random sample of individuals from a given population: $\overline{X} = 172$ cm; $s = 12$ cm; $N = 64$, what would be the approximate probability of drawing a sample ($N = 64$) from that population that would have a mean of *at least* 175 cm? (A) 40% (B) 10% (C) 2% (D) 22%	______	5. _____

(continued)

	Answers	For Scoring
6. For a population with a mean of 200 and a standard deviation of 30, a sample of 100 scores is randomly selected. The probability is 0.50 ($\frac{1}{2}$) that the sample mean falls between what two scores? (A) 198 – 202 (B) 197 – 203 (C) 180 – 220 (D) 194 – 206	______	6. ______
7. The probability of rejecting a null hypothesis that is true is given by: (A) α (B) β (C) standard error of the mean (D) none of the above	______	7. ______
8. The level of significance refers to the risk taken in falsely: (A) rejecting the null hypothesis (B) failing to reject the null hypothesis (C) accepting a type II error (D) making any decision about the hypothesis test	______	8. ______
9. When statistics are used to generalize from a sample to a population, they are referred to as: (A) inferential statistics (B) descriptive statistics (C) graphical statistics (D) normal curve statistics	______	9. ______
10. When a researcher rejects a null hypothesis at the 0.01 level of significance, she is in essence saying, "There is less than 1 chance in 100 that what I have observed in the sample results from the random fluctuations of probability alone." Suppose this rare instance is true in this case and the null hypothesis is rejected erroneously; which of the following statements is true? (A) The researcher has not committed an error. (B) She has committed a type I error. (C) She has committed a type II error. (D) She will be able to quickly determine the reason for rejecting the null hypothesis and to replicate the study to correct the error.	______	10. ______

In the Answers column, write *T* if the statement is true and *F* if the statement is false.

11. The probability of committing a type II error is $1 - \alpha$.	______	11. ______
12. A *z*-score located in the area of rejection leads to the rejection of the null hypothesis.	______	12. ______
13. $(\overline{X} - \mu) \div s_{\overline{X}}$ is an expression equal to a *z*-score.	______	13. ______
14. The absolute value of a *z*-score needed to reject H_0 on a two-tailed test is smaller than the value needed to reject H_0 with a one-tailed test.	______	14. ______
15. α must always be preset to either 0.05 or 0.01.	______	15. ______

Hypothesis Testing

Student ______________________________ Score __________

SECTION 4 PRACTICE PROBLEMS

Items 1 through 5 refer to the problem that follows.

Seniors who work part time are paid on average $\mu = \$6.50$ per hour. Summary data on part-time jobs and wages for the junior class are not available. A random sample $N = 48$ of junior class members who hold part-time jobs revealed $\overline{X} = \$6.15$ per hour with a standard deviation of $1.05.

	Answers	For Scoring
1. What is the standard error of the mean for junior class workers?	__________	1. ____
2. Make an interval estimate of μ for the junior class workers using a 95% confidence interval.	__________	2. ____
3. Use a z-test to help decide whether juniors get paid significantly differently than seniors. Use the 0.05 level of significance and a two-tailed test. Are juniors getting paid significantly differently than seniors?	__________	3. ____
4. Would your decision have been the same if a one-tailed test had been used?	__________	4. ____
5. With α set at 0.05, what is the probability of a type I error in this case?	__________	5. ____

6. Given that IQ scores are distributed throughout the population with a mean of 100 and a standard deviation of 15, what would you conclude about the typical IQ of a class of 36 students at Lincoln High School with a mean class IQ of 104 and a standard deviation of 15? 6. ____

____

____

____

____

____

____

(continued)

For Scoring

7. Over the years, students taking the final exam in American history at Western Heights High School have an average score of 83. Last year's class ($N = 49$) averaged 80 on the test and had a standard deviation of 7.

   a. What is the implied null hypothesis for this problem? 7a. ______

   b. Was last year's class performance inferior to the norm established by prior classes at the 0.05 level of significance on a two-tailed test? 7b. ______

   c. Discuss the probability of a type I and a type II error for this problem. 7c. ______

CHAPTER 12
Correlation

Student ______________________ Total Score ________

SECTION 1 CONCEPT COMPREHENSION

Directions: Complete the following by writing the correct word or words in the respective blanks to the right.

During the physical examination for the basketball team, measures of weight and height were recorded. These data formed a(n) ___(1)___ distribution. To see if these two variables were related, the coach computed a(n) ___(2)___ coefficient and found $r = 0.64$. The plus algebraic sign implied that the taller players tended to weigh ___(3)___ than the shorter players. If these data are considered a sample, the value of r becomes a point estimate of ___(4)___, the population parameter. If the coach wanted to test the null hypothesis that rho = ___(5)___, the degrees of freedom would have to be determined by the formula df = ___(6)___. An interval estimate of rho could be established and would be called a(n) ___(7)___ for rho.

The coach also computed a value of 40.96% (or 0.4096), which provided an indication of how much of the variation in weight would be explained by variation in height. This quantity is called a(n) ___(8)___ of ___(9)___. In examining the relationship between weight and height, the coach had several techniques from which to choose. In this case a Pearson r, or product-___(10)___ r, was computed. Had the players been ranked on both height and weight (instead of measured), a(n) ___*(11)___ rank order correlation could have been used. Because the team has very good overall height, the coach hopes there is a strong ___(12)___ correlation between team height and number of wins in a season.

Answers	For Scoring
1. ____________	1. ____
2. ____________	2. ____
3. ____________	3. ____
4. ____________	4. ____
5. ____________	5. ____
6. ____________	6. ____
7. ____________	7. ____
8. ____________	8. ____
9. ____________	9. ____
10. ____________	10. ____
11. ____________	11. ____
12. ____________	12. ____

*Any item that has an asterisk before the number indicates it covers optional text material. If such material was not presented in class, you may skip this item. Check with your teacher.

SECTION 2 VOCABULARY PRACTICE

Directions: Identify the term defined and complete the designated squares.

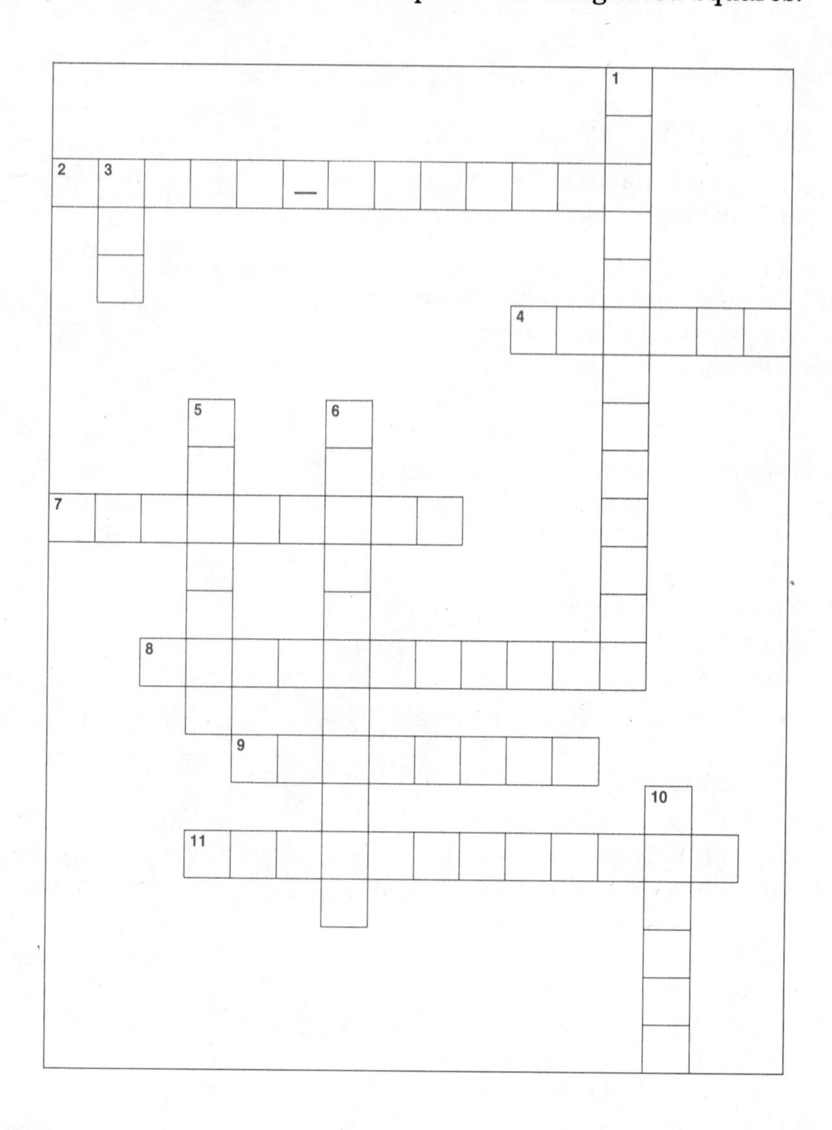

Across

For Scoring

2. XY is the ______ of X and Y. ______
4. Product-______ r. ______
7. Two measures per person forms a ______ distribution. ______
8. r is the symbol for a ______ coefficient. ______
9. r is ______ when high scores on one variable are associated with low scores on the other variable. ______
11. r provides a measure of linear ______ or correlation between two variables. ______

Down

1. r^2 is the coefficient of ______. ______
3. A correlation coefficient parameter. ______
5. Product-moment r is sometimes called ______ r. ______
6. A bivariate graph. ______
10. r is a coefficient of ______ relationship. ______

CHAPTER 12
Correlation

Student ______________________________ Score __________

SECTION 3 REHEARSAL EXERCISES

Directions: In the Answers column, write the letter that represents the correct choice.

	Answers	For Scoring
1. The hypothesis that the value of the correlation coefficient in the population is zero is a(n): (A) alternate hypothesis (B) research hypothesis (C) directional hypothesis (D) null hypothesis	______	1. ______
2. In testing a correlational null hypothesis, the *critical* (tabled) *value* of r decreases as N: (A) decreases (B) increases (C) remains constant (D) approaches zero	______	2. ______
3. If $r = +0.80$, which of the following 95% confidence intervals for rho is most plausible? (A) {0.60 to 0.90} (B) {0.80 to 0.90} (C) {0.80 to 1.30} (D) {0.50 to 0.80}	______	3. ______
4. Given that rho, the population correlation coefficient, is zero, what is the probability that a sample ($N = 32$) bivariate distribution with an $r = +0.296$ or larger will be drawn by chance? (A) 0.10 (B) 0.05 (C) 0.025 (D) 0.30	______	4. ______
5. If 37 students are given both a music test and a physics test, and if $r = 0.32$ is computed, the degrees of freedom for the hypothesis test equal: (A) 37 (B) 32 (C) 30 (D) 35	______	5. ______
*6. Spearman's correlation should be used when: (A) the Pearson r is too difficult to compute (B) the statistician wants to know how curvilinear a relationship is (C) the bivariate distribution data are measured on an ordinal scale (D) the original data are measured on a ratio scale	______	6. ______

(continued)

Answers | For Scoring

7. As the ellipse formed by points on a scattergram becomes thinner, the correlation between the scores: ______ 7. ______
(A) becomes lower
(B) becomes higher
(C) remains constant
(D) becomes curvilinear

8. The lowest magnitude (strength) of correlation among the following is: ______ 8. ______
(A) 0.26
(B) −0.45
(C) 0.05
(D) 0.90

9. Imagine measuring the respective distances from the ground of two boys on a seesaw as they move up and down 50 times. The Pearson r between the heights would be: ______ 9. ______
(A) 0.00
(B) +1.00
(C) −1.00
(D) 0.50

10. If the correlation between X and Y is −0.94, which of the following statements is true? ______ 10. ______
(A) X causes Y.
(B) Y causes X.
(C) Low scores on X are associated with high scores on Y.
(D) Low scores on X are associated with low scores on Y.

In the Answers column, write T if the statement is true and F if the statement is false.

11. r^2 is called the coefficient of determination. ______ 11. ______

12. If $r = -1.00$, a raw score value of 39 on one variable in the bivariate distribution would have to be offset by a value of −39 on the other variable. ______ 12. ______

13. The range of a correlation coefficient is 0 to +1. ______ 13. ______

*14. Spearman can be used to compute correlations between nominal variables. ______ 14. ______

15. The degrees of freedom for a Pearson r decrease as N becomes smaller. ______ 15. ______

16. A researcher obtained a correlation $r = +0.84$ between the amount of time spent watching television and the number of pounds the viewers were over their recommended weight. This implies that overweight people in that population tend to watch more TV than normal-weight people. ______ 16. ______

17. The value Σxy can never be negative. ______ 17. ______

(continued)

Student ______________________ Score ________

SECTION 3 REHEARSAL EXERCISES (continued)

	Answers	For Scoring
18. If an r is statistically significant at the 0.01 level, it is also significant at the 0.05 level.	______	18. ______
19. If $\Sigma xy = 0$ for a bivariate distribution, then $r = 0$ for the data.	______	19. ______
20. If all the points of a scattergram form a straight line, then $r = 0.00$.	______	20. ______

SECTION 4 PRACTICE PROBLEMS

Perform the indicated calculations.

	Answers	For Scoring
1. If $r = 0.67$ and $N = 10$, compute the 95% confidence interval for rho.	______	1. ______
2. Given that $N = 52$ and $r = 0.62$, find the 95% confidence interval for rho.	______	2. ______
3. Given that $r = -0.9$, find the coefficient of determination.	______	3. ______
4. Given that $\Sigma X = 21$, $\Sigma Y = 29$, $\Sigma XY = 112$, $\Sigma X^2 = 91$, and $\Sigma Y^2 = 151$, find r. $N = 6$ for the problem.	______	4. ______
5. What two values of r could result in a coefficient of determination of 0.49?	______	5. ______

6. Given the following scores of 15 students on an algebra exam and a chemistry exam, a. compute a Pearson r and *b. rank order the scores for both exams and calculate the Spearman's correlation.

Student ID Number	Algebra Score	Chemistry Score
1	92	90
2	85	71
3	84	73
4	83	95
5	79	83
6	77	80
7	76	80
8	75	77
9	74	64
10	73	72
11	68	65
12	66	60
13	62	68
14	52	63
15	51	55

	Answers	For Scoring
a. Pearson r = ______	______	6a. ______
*b. Spearman = ______	______	6b. ______

(continued)

Answers | For Scoring

7. Given the data that follow, complete the indicated summary calculations and correlational analyses.

	ID Number	X	X^2	Y	Y^2	XY
a.	01	7	______	3	______	______
	02	13	______	6	______	______
	03	2	______	2	______	______
	04	4	______	5	______	______
	05	15	______	14	______	______
	06	10	______	10	______	______
	07	19	______	8	______	______
	08	28	______	19	______	______
	09	26	______	15	______	______
	10	22	______	17	______	______
		______ ΣX	______ ΣX^2	______ ΣY	______ ΣY^2	______ ΣXY

7a. ______

b. $\Sigma x^2 = \Sigma X^2 - \frac{(\Sigma X)^2}{N} =$ ______ 7b. ______

$\Sigma y^2 = \Sigma Y^2 - \frac{(\Sigma Y)^2}{N} =$ ______

$\Sigma xy = \Sigma XY - \frac{(\Sigma X)(\Sigma Y)}{N} =$ ______

c. Find r: $r = \frac{\Sigma xy}{\sqrt{(\Sigma x^2)(\Sigma y^2)}}$ ______ 7c. ______

d. State the null hypothesis for the correlational problem. 7d. ______

____

____

(continued)

Student ______________________ Score __________

SECTION 4 PRACTICE PROBLEMS (continued)

For Scoring

e. What is the result of a hypothesis test at the 0.05 level of significance using a two-tailed test? 7e. ____

____

____

f. Plot a bivariate graph (scattergram) of the data. 7f. ____

Y: 0, 5, 10, 15, 20

X: 5, 10, 15, 20, 25

	Answers	For Scoring
g. (1) What is the value of z_r?	________	7g.(1) ____
(2) What is the standard error of z_r?	________	7g.(2) ____
(3) What is the 95% confidence interval?	________	7g.(3) ____
(4) What is the coefficient of determination?	________	7g.(4) ____

(continued)

For Scoring

8. Eight students in the high-school ROTC cadet program took a 10-item written test (Y) on physical fitness. They also were asked to count the number of push-ups they could do in a 30-second time frame (X).

8. ______

X	Y
17	8
13	6
15	7
15	6
12	4
14	5
14	5
10	3

Is there a statistically significant relationship between X and Y? Justify your response.

9. Given the bivariate distribution of X and Y:

X	Y
20	5
16	8
10	12
12	9
14	9
17	5
15	10
10	14
8	15

a. State the implied null hypothesis for a correlation problem.

9a. ______

b. Test H_0 at the 0.01 (two-tailed) level of significance.

9b. ______

CHAPTER 13
Regression Analysis

Student ______________________________ Total Score __________

SECTION 1 CONCEPT COMPREHENSION

Directions: Complete the following by writing the correct word or words in the respective blanks to the right.

The purpose of ___(1)___ is to quantify the magnitude and direction of a relationship between two variables, whereas ___(2)___ procedures are used for predicting. If the strength of a relationship is sufficient, one variable known as the ___(3)___ variable is used as a predictor of another variable called the ___(4)___ variable. The stronger the relationship, the ___(5)___ accurate predictions can be. However, because correlations are virtually never perfect in the real world, error will be present to some extent and is measured by the ___(6)___ of ___(7)___.

In Chapter 13, *X* is used as the independent variable to "explain" the variability in *Y*. The variability in *Y* as measured by the sum of squares of the deviation scores can be conceptualized as having two sources: (1) regression and (2) ___(8)___. To predict values of the dependent variable, a regression equation is necessary. The equation defines a straight ___(9)___ of prediction and consists of two constants. One constant, symbolized *a*, is the ___(10)___ of the *Y*-axis, whereas the other constant is called the ___(11)___ coefficient. The slope of the line, when taken as a population parameter, is called ___(12)___ and can be tested for statistical significance using a(n) ___(13)___-test with two sets of degrees of ___(14)___.

The regression line is known as the line of best fit or the line of least ___(15)___, because the squares of the errors are minimized. The standard deviation of the errors in prediction, that is, the difference between *Y* and *Y′*, is known as the standard ___(16)___ of estimate. A region about the regression line that includes 95% of the bivariate points is known as a 95% confidence ___(17)___. In summary, the accuracy of ___(18)___ is enhanced if the relationship between the independent and dependent variables is strong. If the linear relationship is weak, the standard error of estimate will be relatively ___(19)___, which means that predictions (*Y′*) generally will be poor estimates of actual *Y*-values.

Answers	For Scoring
1. ____________	1. ____
2. ____________	2. ____
3. ____________	3. ____
4. ____________	4. ____
5. ____________	5. ____
6. ____________	6. ____
7. ____________	7. ____
8. ____________	8. ____
9. ____________	9. ____
10. ____________	10. ____
11. ____________	11. ____
12. ____________	12. ____
13. ____________	13. ____
14. ____________	14. ____
15. ____________	15. ____
16. ____________	16. ____
17. ____________	17. ____
18. ____________	18. ____
19. ____________	19. ____

SECTION 2 VOCABULARY PRACTICE

Directions: Identify the term defined and complete the designated squares.

Across — For Scoring

2. b is the symbol for a regression ______. ______
6. Means equal scatter of the bivariate points along the regression line. ______
7. Means a distribution along a straight line. ______
8. Abbreviation for degrees of freedom. ______
9. The regression coefficient parameter for a population. ______

Down

1. A linear method of prediction. ______
3. Quantified by the regression coefficient. ______
4. Error in prediction is measured by the standard error of ______. ______
5. The variable being predicted in a linear regression. ______
10. The point at which the regression line crosses the axis of the dependent variable is called a(n) ______. ______

Regression Analysis

Student ______________________________ Score __________

SECTION 3 REHEARSAL EXERCISES

Directions: In the Answers column, write the letter that represents the correct choice.

Answers | For Scoring

1. The intercept of $Y' = 5 - 2X$ is: ______ 1. ______
   (A) 5
   (B) 2
   (C) −2
   (D) $-\frac{2}{5}$

2. In the linear regression equation for predicting Y from X, if $b = 0$, then: ______ 2. ______
   (A) the prediction will be without error
   (B) there is no relationship between X and Y
   (C) both (A) and (B) are true
   (D) none of the above is true

3. The regression coefficient (b) for a bivariate regression analysis is the: ______ 3. ______
   (A) point at which the regression line crosses the axis of the independent variable
   (B) point at which the regression line crosses the axis of the dependent variable
   (C) slope of the regression line
   (D) standard error of estimate

4. A college admissions counselor found that an aptitude test (X) correlated 0.94 with GPA in freshman courses at the nearby university. The standard deviation of the aptitude test was 16 points, the standard error of estimate for predicting GPA (Y) was 0.1, and the regression equation was $Y' = 0.6 + 0.02X$. Based on these results, if a student scored 110 on the aptitude test, about what are her/his chances of making *at least* a 2.75 GPA in freshman courses at the university? ______ 4. ______
   (A) 0.69
   (B) 0.31
   (C) 0.05
   (D) 0.56

5. Given: $N = 41$, $\overline{Y} = 10$, $\overline{X} = 12$, $\Sigma(X - \overline{X})^2 = 400$, $\Sigma(Y - \overline{Y})^2 = 64$, and $\Sigma(X - \overline{X})(Y - \overline{Y}) = 100$. What is the predicted value of Y given that X is 8? ______ 5. ______
   (A) 6.4
   (B) 62.5
   (C) 4.75
   (D) 9.0

(continued)

SECTION 3 REHEARSAL EXERCISES (continued)

Answers **For Scoring**

6. You can sketch a fairly accurate line of linear regression if you are provided only: ______ 6. ______
   (A) the slope and intercept of the regression line
   (B) the standard error of estimate and the slope
   (C) the mean of X and the mean of Y
   (D) the intercept of the regression line and the bivariate coordinates of one case from the sample

7. For a regression problem, if $X = 2$, $a = 3$, and $b = 0.5$, then Y' (the predicted value of Y) equals: ______ 7. ______
   (A) 3
   (B) 6.5
   (C) 4
   (D) 2.5

8. Analogy: "correlation" is to "regression" as "relationship" is to: ______ 8. ______
   (A) scattergram
   (B) prediction
   (C) sample
   (D) slope

9. In regression, a statistical test of the null hypothesis that $\beta = 0$ is: ______ 9. ______
   (A) redundant to a test of H_0: $\rho = 0$
   (B) accomplished with an F-test
   (C) a test that the slope is zero in the population
   (D) all of the above

10. The coefficient of determination, r^2, can be found by: ______ 10. ______
   (A) $\dfrac{SS_{regression}}{SS_{error}}$
   (B) $SS_r - SS_e$
   (C) $MS_r \div MS_e$
   (D) $\dfrac{SS_{regression}}{SS_{total}}$

SECTION 4 PRACTICE PROBLEMS

For problems 1 through 10, use the following data, perform the indicated calculations, and report your answers to two decimal places.

$N = 18$; $\Sigma X = 82$; $\Sigma Y = 76$; $\Sigma X^2 = 450$; $\Sigma Y^2 = 375$;
$\Sigma XY = 390$; $\overline{X} = 4.56$; $\overline{Y} = 4.22$

Answers **For Scoring**

1. Find Σx^2. ______________ 1. ______

2. Find Σy^2. ______________ 2. ______

3. Find Σxy. ______________ 3. ______

4. What is the value of r? ______________ 4. ______

(continued)

Student ______________________________ Score __________

SECTION 4 PRACTICE PROBLEMS (continued)

Answers **For Scoring**

5. What is the coefficient of determination? __________ 5. ____

6. Compute the value of b for regression. __________ 6. ____

7. Compute the value of a for regression. __________ 7. ____

8. Write out the regression equation. __________ 8. ____

9. If $X = 4$, $Y' =$ ______. __________ 9. ____

10. Calculate the value of the standard error of estimate. __________ 10. ____

11. Use the following data to complete the worksheet specifications for parts a. through d.

a. 11a. ____

	X	Y	X^2	Y^2	XY	Y'
	4	3	____	____	____	____
	6	10	____	____	____	____
	9	5	____	____	____	____
	11	23	____	____	____	____
	15	25	____	____	____	____
Sums	____	____	____	____	____	____
	ΣX	ΣY	ΣX^2	ΣY^2	ΣXY	$\Sigma Y'$

b. Perform the necessary computations to determine the regression equation for the data. 11b. ____

$$Y' = \_\_\_\_\_\_ + (\_\_\_\_\_\_)X$$

Use the equation to complete the right-hand column in the table—predict Y for each value of X in the bivariate set.

c. Find the sum of squares of the values in the Y' column; that is, treat Y' as a variable and compute the sum of squares. Call it $\Sigma y'^2 =$ __________. 11c. ____

d. Compare Σy^2 with $\Sigma y'^2$ by dividing the latter by the former. This quotient is the proportion of variability (as measured by sum of squares) in one variable that can be accounted for by variability (as measured by sum of squares) in the other variable. What other measure that has been discussed provides the same value? 11d. ____

____

____

(continued)

12. The Home and Family Living class at Fair View High School collected bivariate data on the average monthly temperature (X) and the average electric bill (Y) for the households represented by the class members. The data were:

Month	X Average Temperature (°F)	Y Average Bill ($)
January	25	76
February	33	85
March	50	92
April	54	90
May	65	104
June	80	110
July	90	120
August	95	115
September	75	108
October	60	97
November	45	92
December	34	86

a. Assuming a linear relationship exists between X and Y, what is the least-squares regression line for predicting the average electric bill from the average monthly temperature? __________ 12a. ____

b. Suppose an almanac forecasts an 8° increase in temperature for the upcoming month of June over the previous June's temperature. Assume that the cost of electricity will rise 10% over the previous year because of inflation alone and determine how much you would recommend budgeting for the average June electric bill. __________ 12b. ____

c. What is the standard error of estimate for predicting the monthly electric bill? __________ 12c. ____

(continued)

Student ______________________________ Score ____________

SECTION 4 PRACTICE PROBLEMS (continued)

Answers **For Scoring**

13. Fifteen students were assessed with a social personality instrument. Their scores and high school grade-point averages are as follows:

Personality	High-School GPA
70	3.8
63	3.7
32	1.9
50	3.0
45	2.8
20	2.2
30	2.4
30	2.3
50	2.8
72	3.9
56	3.4
47	2.9
50	3.3
60	3.5
42	3.0

a. What is the Pearson correlation coefficient? ____________ 13a. _____

b. Is the relationship statistically significant at the 0.05 level of significance? ____________ 13b. _____

c. Compute a. and b. for the least-squares regression line. What is the regression equation for predicting GPA from personality? ____________ 13c. _____

d. What would a student's predicted GPA be if he or she scored 55 on the personality scale? ____________ 13d. _____

e. What is the standard error of estimate in predicting high school GPA from personality? ____________ 13e. _____

f. What is the probability that a student who scores 55 on the personality scale will earn a high school GPA of 3.0 or higher? ____________ 13f. _____

(continued)

14. Describe the influence of sample size (N) on the standard error of estimate. (Hint: Examine the influence of N in the computational formula.) 14. ______

______

______

______

15. A statistics teacher at Springdale High School claims that scores on the first examination are very accurate predictors of final grades in the course. To test this conjecture, the teacher recorded exam scores and final averages for 15 students, as follows:

Student	First Exam	Final Average
01	65	61
02	85	93
03	73	90
04	85	91
05	97	95
06	77	72
07	89	84
08	70	69
09	99	96
10	77	83
11	90	86
12	75	80
13	72	74
14	70	67
15	80	80

a. Is the teacher's assertion correct? Test at the 0.01 level of significance. 15a. ______

______

______

______

Answers

b. What is the equation of the least-squares regression line? ______ 15b. ______

c. Compute the standard error of estimate for predicting the final average. ______ 15c. ______

d. If this relationship remained the same for a 10-year period and if during that time span there were 60 students who scored 77 on the first exam, how many of these 60 would be expected to average 80 or higher on the final? ______ 15d. ______

(continued)

Student ______________________________ Score __________

SECTION 4 PRACTICE PROBLEMS (continued)

For Scoring

e. On the grid that follows, label and scale the axes and sketch the regression line and a 68% and 95% confidence band.

15e. _____

t-Test

Student ______________________ Total Score __________

SECTION 1 CONCEPT COMPREHENSION

Directions: Complete the following by writing the correct word or words in the respective blanks to the right.

The *t*-test is used to test for significant differences between (1) of two sets of data. If one group of subjects have been assessed two times to create the two sets of data, the (2) *t*-test model is appropriate. On the other hand, if two different groups of subjects have been measured with one instrument on one occasion, a(n) (3) *t*-test model is warranted. If the data are from two distinct groups, the *F*-test is used to test for homogeneity of (4). If the null hypothesis that the variances of the two sets of data are equal is found tenable, then the (5) variance *t*-test formula is best for providing a test difference between the means. Further, if the two groups have significantly different variances and have different *N*'s, the (6) variance *t*-test formula is recommended.

With all models of the *t*-test, the calculated value of *t* is compared to a(n) (7) (or tabled) value of *t* to determine if the null hypothesis should be rejected. If the calculated value of *t* is (8) in absolute terms than the tabled value of *t*, the results are deemed statistically significant. However, before one can determine the critical value of *t*, the (9) must be determined. In the case of the pooled variance model, df = [(10)]. As will all inferential hypothesis tests, the probability of committing a type I error if the null hypothesis is incorrectly rejected is the preset (11) (or significance level).

Several factors influence the magnitude of the calculated *t*-value. For example, the (12) the difference between the means, the larger the *t*-value. If all else is held constant, the (13) the within-group variance, the smaller the *t*-value, whereas the (14) the sample sizes, the larger the calculated value of *t*. If a directional prediction about which of two means will be larger (if a significant difference exists) can be justified, then a (15)-tailed test is appropriate. Finally, the (16) *t*-test model will give identical results to the (17) model if $r = 0$.

	Answers	For Scoring
1.		1.
2.		2.
3.		3.
4.		4.
5.		5.
6.		6.
7.		7.
8.		8.
9.		9.
10.		10.
11.		11.
12.		12.
13.		13.
14.		14.
15.		15.
16.		16.
17.		17.

SECTION 2 VOCABULARY PRACTICE

Directions: Identify the term defined and complete the designated squares.

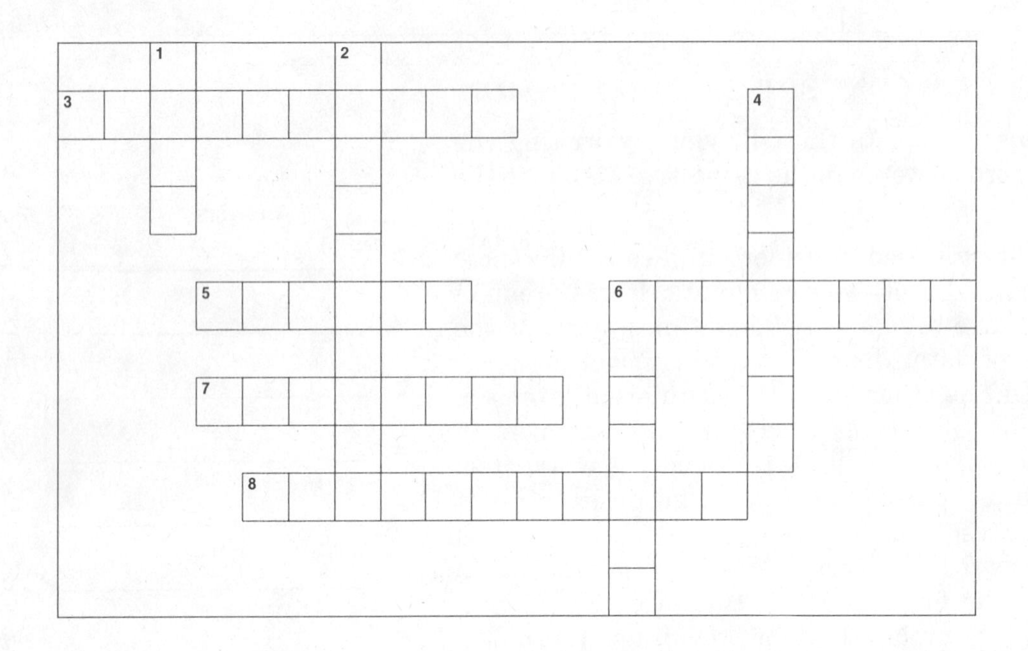

Across

For Scoring

3. The *t*-test model to use when one group of subjects have been measured twice. ________
5. If two groups do not have significantly different variances, the differences in their means may be tested with the ______ variance model. ________
6. If two groups have significantly different variances and are unequal in size (*N*'s), the ______ variance is the appropriate *t*-test model. ________
7. The tabled value of *t* is the absolute value of the ______ *t*-value. ________
8. The pooled and separate variance *t*-models are ______ *t*-tests. ________

Down

1. The ______ of rejection for a one-tailed test is located in the extreme portion of the *t*-distribution. ________
2. If the ______ value of *t* is less (in absolute terms) than the tabled value of *t*, H_0 is considered tenable. ________
4. The *F*-test is used to test for homogeneity of ______. ________
6. Gossett's status when he discovered the *t*-distribution. ________

CHAPTER 14
t-Test

Student ______________________________ Score __________

SECTION 3 REHEARSAL EXERCISES

Directions: In the Answers column, write the letter that represents the correct choice.

	Answers	For Scoring
1. If 31 students are used in a correlated *t*-test design, the degrees of freedom equal: (A) 31 (B) 30 (C) 29 (D) 60	______	1. _____
2. The areas of rejection for a two-tailed test of significance using a *t*-test are: (A) between the two tails of the *t*-distribution (B) in the two extreme tail sections of the *t*-distribution (C) in one tail of the *t*-distribution (D) between the critical values of *t*	______	2. _____
3. If the calculated *t*-value is larger than the tabled *t*-value, the null hypothesis is: (A) tenable (B) impossible (C) not rejected (D) improbable	______	3. _____
4. To test for a significant difference between the variances of two sets of data, the appropriate statistic is: (A) a correlated *t*-test (B) an independent *t*-test (C) *r*: a correlation coefficient (D) an *F*-test	______	4. _____
5. The null hypothesis is a statement of an expected outcome under: (A) experimental conditions (B) normal conditions (C) random conditions (D) nonrandom conditions	______	5. _____
6. Which of the following statements does not belong with the other three? (A) The calculated *t*-value is less than the tabled *t*-value. (B) The null hypothesis was considered tenable. (C) The difference between the sample means was a result of sampling error and chance. (D) A statistically significant difference was found between the means.	______	6. _____

(continued)

SECTION 3 REHEARSAL EXERCISES (continued)

	Answers	For Scoring
7. A researcher who measured the same group of students under two experimental conditions erroneously used a separate variance *t*-test to test the null hypothesis. Upon discovering her error, she recomputed the *t*-value using the correlated *t*-test model. The new (and correct) *t*-value will be: (A) smaller under any circumstance (B) smaller only if $r > 0$ (C) larger under any circumstance (D) larger if $r > 0$	______	7. _____
8. If the calculated value of t is -2.9 on a one-tailed test at the 0.05 level of significance with eight degrees of freedom, then one should: (A) not reject H_0 (B) reject H_0 (C) recalculate, because t cannot be negative (D) accept H_0	______	8. _____
9. What is the critical value of t for a two-tailed test at the 0.05 level of significance for a separate variance model in which the group sizes are 21 and 16, respectively? (A) 2.086 (B) 2.131 (C) 2.1085 (D) 2.04	______	9. _____
10. As the degrees of freedom increase, the critical *t*-values: (A) decrease (B) remain constant (C) increase (D) approach zero	______	10. _____

SECTION 4 PRACTICE PROBLEMS

Use the table of *t*-values to respond to problems 1 and 2.

	Answers	For Scoring
1. What proportion of the area of the *t*-distribution falls:		
a. above $t = 2.947$ when df = 15?	__________	1a. _____
b. below $t = -1.812$ when df = 10?	__________	1b. _____
c. above $t = 2.228$ when df = 10?	__________	1c. _____
d. between $t = \pm 1.96$ when df is infinite?	__________	1d. _____
2. Find the value of t for df = 25 such that the proportion of the area:		
a. to the left of t is 0.025	__________	2a. _____
b. to the right of t is 0.005	__________	2b. _____
c. between the mean and t is 0.45	__________	2c. _____
d. between $-t$ and $+t$ is 0.95	__________	2d. _____

(continued)

Student ______________________ Score __________

SECTION 4 PRACTICE PROBLEMS (continued)

For problems 3, 4, and 5, determine whether the separate variance or the pooled variance *t*-test model should be used and specify the degrees of freedom in each instance.

Answers | **For Scoring**

3. Group 1: $N = 25$; $\Sigma x^2 = 2{,}400$; $\overline{X} = 62.80$ Model: __________ 3. ____
   Group 2: $N = 14$; $\Sigma x^2 = 1{,}200$; $\overline{X} = 73.42$
   df = __________

4. Group 1: $N = 14$; $\Sigma x^2 = 1{,}327$; $\overline{X} = 43.71$ Model: __________ 4. ____
   Group 2: $N = 14$; $\Sigma x^2 = 260$; $\overline{X} = 36.28$
   df = __________

5. Group 1: $N = 24$; $\Sigma x^2 = 629$; $\overline{X} = 106.12$ Model: __________ 5. ____
   Group 2: $N = 16$; $\Sigma x^2 = 1{,}607$; $\overline{X} = 100.22$
   df = __________

For problems 6a. through f., compute the necessary summary statistics, determine the correct *t*-model, and test the null hypothesis for statistical significance at the 0.05 level using a two-tailed test.

6. In an experiment, one group was treated with a placebo while another was treated with an experimental manipulation. The data are as follows:

Placebo (Group P)	Experimental (Group E)
26	38
24	26
18	24
17	24
18	30
20	22
18	

a. State the null hypothesis. 6a. ____

____

____

b. *t*-test model used: __________ 6b. ____

c. Calculated *t*-value: __________ 6c. ____

d. Degrees of freedom for the problem: __________ 6d. ____

e. Critical (tabled) *t*-value: __________ 6e. ____

f. Do the two treatments appear to differentiate in the scores more than would be expected just by chance? __________ 6f. ____

(continued)

 Answers **For Scoring**

7. Fifteen students were given an achievement test. The test consisted of two parts. One was a verbal measure of achievement (Verbal); the other was more number oriented and was called Quantitative achievement. The scores are as follows:

Verbal	Quantitative	Verbal	Quantitative
60	55	58	54
80	76	90	83
91	90	85	79
53	50	43	40
47	41	59	62
65	65	45	47
50	52	92	88
75	70		

a. Complete the summary calculations: 7a. _____

	Verbal	Quantitative		
$\overline{X}$	__________	__________		
s^2	__________	__________		
s	__________	__________	$r =$	__________

b. Calculated t-value: __________ 7b. _____

c. Critical t-value: __________ 7c. _____

d. What conclusion about the relative performance on verbal and quantitative achievement can be drawn based on the analysis? 7d. _____

____

____

____

____

8. Two randomly assigned groups of students took a geography test. One group had been taught by lecture and photographic slides while the other group had been taught in a map-reading laboratory setting. Test results were: 8. _____

Lecture Group	Laboratory Group
15	12
14	17
12	11
15	19
13	13
12	13
16	16
14	19
10	12
13	16

(continued)

Student ______________________________ Score __________

SECTION 4 PRACTICE PROBLEMS (continued)

For Scoring

Appropriately test the null hypothesis at the 0.05 level of significance with a two-tailed test and state your conclusions.

____

____

____

____

____

9. A statistics problem is solved by a group of students using two different brands of calculator. Half the students used brand T first and brand S second; the other half used S, then T for solving the problem to counterbalance any effect of practice. The times (in seconds) required for the students to perform the computations were:

9. _____

	Calculator	
Student	**Brand T**	**Brand S**
C.A.	50	57
R.B.	32	32
J.C.	42	48
K.E.	71	68
J.F.	67	70
B.G.	50	62
D.J.	54	51
J.J.	35	35
L.M.	50	58
A.S.	42	63
P.S.	60	58
B.W.	35	43

Do these data provide evidence at the 0.05 level of significance that one brand of calculator is more efficient for such calculations? Why or why not?

____

____

____

____

____

____

(continued)

For Scoring

10. The statistics class at Sandia Academy tested two brands of calculator batteries. A complicated scoring scheme was developed that included the life of the battery, the cost, its effectiveness after recharging, and its durability. Scores on samples of two brands are:

Brand A	Brand B
10	7
3	1
5	9
7	5
2	3
3	9
5	4
6	2
	5

Use the appropriate *t*-test model and decide if one brand is superior to the other at the 0.01 level of significance. State your results and conclusion, and justify your decision.

10. _____

CHAPTER 15
Analysis of Variance

Student ______________________________ Total Score ____________

SECTION 1 CONCEPT COMPREHENSION

Directions: Complete the following by writing the correct word or words in the respective blanks to the right.

Analysis of variance (ANOVA) is a popular technique for testing differences between group __(1)__ to determine if they are statistically significant. A(n) __(2)__ statistic is used to compare two components of variance. This distribution, unlike the normal and *t*-distributions, is not symmetrical; it is __(3)__ skewed. The area of rejection is in the tail in the extreme __(4)__ or upper portion of the distribution. When the calculated value of F exceeds the __(5)__ or tabled value, the null hypothesis is __(6)__.

The rationale for ANOVA rests on comparing two components of variance. One component of variance is error variance or __(7)__-group variance. Another component is treatment or __(8)__-groups variance. For each source of variance, the sums of squares are divided by the __(9)__ to obtain the __(10)__ square values. If the mean square __(11)__ groups is much larger than the mean square __(12)__ a group, the difference among the means will probably be statistically significant.

In an ANOVA problem, if there are more than two groups ($k > 2$) and F is significant, then multiple comparison tests are probably appropriate. These are also referred to as *a posteriori* or __(13)__ tests. Their purpose is to identify pairwise differences after ANOVA has indicated an overall difference. If there are k groups, the number of possible pairwise comparisons is __(14)__. A very widely used *a posteriori* test was discussed in the text; it involves computing an F-value for each combination of means and is called the __(15)__ test.

	Answers	For Scoring
1.	____________	1. ____
2.	____________	2. ____
3.	____________	3. ____
4.	____________	4. ____
5.	____________	5. ____
6.	____________	6. ____
7.	____________	7. ____
8.	____________	8. ____
9.	____________	9. ____
10.	____________	10. ____
11.	____________	11. ____
12.	____________	12. ____
13.	____________	13. ____
14.	____________	14. ____
15.	____________	15. ____

SECTION 2 VOCABULARY PRACTICE

Directions: Identify the term defined and complete the designated squares.

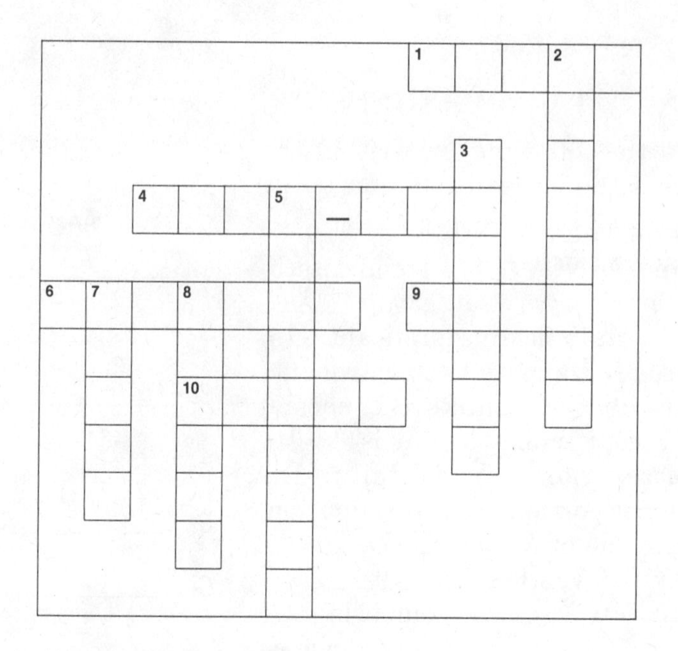

Across **For Scoring**

1. Acronym for analysis of variance. ________
4. Another name for *a posteriori*. ________
6. ______ variance is caused by differences among group means. ________
9. ______ square means variance. ________
10. Within-group df + between-groups df = ______ df. ________

Down

2. Sum of squares divided by degrees of freedom = ______. ________
3. A popular *a posteriori* test. ________
5. ______ variance is the same as between-groups variance. ________
7. Within-group variance is called ______ variance. ________
8. Individual differences that cannot be explained by differential treatment or by group membership. ________

CHAPTER 15

Analysis of Variance

Student ______________________________ Score __________

SECTION 3 REHEARSAL EXERCISES

Directions: In the Answers column, write the letter that represents the correct choice.

Answers **For Scoring**

1. If an ANOVA revealed significant differences among seven groups, how many pairwise multiple comparisons are possible? ______ 1. ______
   (A) 42
   (B) 7
   (C) 21
   (D) none of the above
2. "Mean squares" means: ______ 2. ______
   (A) sum of squares
   (B) analysis of variance
   (C) variance
   (D) between groups
3. The area of rejection in an *F*-distribution is in: ______ 3. ______
   (A) both tails
   (B) the left tail
   (C) the right tail
   (D) the central portion
4. Given that all other factors are held constant, the larger the MS within a group, the: ______ 4. ______
   (A) smaller the *F*-value
   (B) larger the *F*-value
   (C) smaller the MS between groups
   (D) more likely the null hypothesis will be rejected
5. As variability among means increases, so does: ______ 5. ______
   (A) the *F*-ratio
   (B) the within-group variance
   (C) the error variance
   (D) all of the above
6. Suppose that 10 groups are being compared with ANOVA and that $SS_w = 18$ and $SS_b = 16$. What is the between-groups mean square? ______ 6. ______
   (A) 2.00
   (B) 3.40
   (C) 1.78
   (D) none of the above
7. *Post hoc* tests are used: ______ 7. ______
   (A) before ANOVA to determine if *F* is significant
   (B) when two groups are analyzed by ANOVA
   (C) only if the null hypothesis is not rejected, to find out why
   (D) after ANOVA when the null hypothesis that three or more means are equal is rejected

(continued)

SECTION 3 REHEARSAL EXERCISES (continued)

	Answers	For Scoring

8. If three groups were analyzed in an ANOVA with 10 in each group, the degrees of freedom for the F-ratio value would equal: ______ 8. ______
   (A) 2 and 27
   (B) 3 and 30
   (C) 2 and 3
   (D) 3 and 29
9. Which of the following increases as the differences among group means increase? ______ 9. ______
   (A) sum of squares between groups
   (B) mean square between groups
   (C) calculated F-ratio
   (D) all of the above
10. If an F-ratio fails to reject the null hypothesis at the 0.05 level, the researcher: ______ 10. ______
    (A) runs a 5% risk of having committed a type I error
    (B) has committed a type II error
    (C) has not committed an error
    (D) runs a risk of having committed a type II error

SECTION 4 PRACTICE PROBLEMS

Solve the following ANOVA problem and respond to problems 1 through 3.

A sample of seven male and seven female students from the communications club were asked how many hours of TV they regularly watched during the week. The data are as follows:

Male	Female
14	18
23	22
19	22
10	12
16	19
12	25
20	23

For Scoring

1. Complete the ANOVA summary table that follows: 1. ______

Source	DF	SS	MS	F
Between				
Within				
Total				

2. Complete the table of descriptive statistics: 2. ______

	$\overline{X}$	s
Male		
Female		

(continued)

Student ______________________ Score __________

SECTION 4 PRACTICE PROBLEMS (continued)

For Scoring

3. Using $\alpha = 0.05$, interpret the results. 3. ____

____

____

____

____

____

Solve the following ANOVA problem and respond to problems 4 through 7.

For an entire year, library usage required for various classes was recorded. The number of hours per month of required library work for three classes are as follows:

Literature	History	Biology
18	15	10
10	10	12
16	12	10
19	14	9
15	10	12
17	11	15
15	14	11
20		10
17		

4. Complete the ANOVA summary table: 4. ____

Source	DF	SS	MS	F
Between				
Within				
Total				

5. Calculate the summary descriptive statistics indicated below: 5. ____

	$\overline{X}$	s
Literature		
History		
Biology		

(continued)

For Scoring

6. Perform Scheffé's *a posteriori* test and summarize the results below: 6. ______

Pair	F-ratio	Significance 0.01 0.05 ns
Literature vs. History		
Literature vs. Biology		
History vs. Biology		

7. Interpret the preceding results in light of the 0.05 level of significance—tell where (between which groups) the differences exist (if at all) and where differences do not exist (if applicable). 7. ______

____

____

____

____

____

8. The data from Practice Problem 8 in Chapter 14 are repeated here. Perform an ANOVA on the data and compare the conclusions with your conclusions from the *t*-test results.

Lecture Group	Laboratory Group
15	12
14	17
12	11
15	19
13	13
12	13
16	16
14	19
10	12
13	16

a. ANOVA results: _______________ 8a. ______

____

____

____

b. Comparison to *t*-test: _______________ 8b. ______

____

____

____

(continued)

Student ______________________ Score __________

SECTION 4 PRACTICE PROBLEMS (continued)

For Scoring

9. The sophomore class of Middleton High School is going to sell T-shirts as a money-raising activity. Using a 20-point scale (1 = poor to 20 = excellent), the class marketing committee rated three slogans that were being considered as imprints on the T-shirts. The ratings were as follows:

Slogan A	Slogan B	Slogan C
18	17	15
12	19	12
14	18	10
10	18	12
8	15	13
15	12	6
13	9	10
13	15	7
17	19	6

a. Determine if there are differences in the ratings of the T-shirt slogans at the 0.05 level of significance and, if so, where the differences lie. 9a. ______

b. What advice would you give the class, assuming all factors other than the ratings are equal? 9b. ______

(continued)

For Scoring

10. Students in the sociology club at Maderno High School devised an experiment to test the effect of appearance on behavior. A team of students randomly selected 40 households near the school campus to participate in an interview about the quality of the school. In half of the interviews (randomly determined), the team dressed up in business suits; in the other half, the team dressed casually in cut-offs and tattered tennis shoes. The ratings given to the school by the two groups of adults were:

10. ______

Dressed-up Group				Casually Dressed Group			
17	13	18	19	15	10	17	16
12	18	12	17	14	16	13	18
15	19	14	12	12	15	16	14
15	16	15	15	15	18	15	15
14	15	13	16	13	18	12	14

Use ANOVA at the 0.05 level of significance to determine if the household interviewees responded differently to the teams about the quality of the school.

______

______

______

______

______

______

CHAPTER 16
Chi Square

Student ______________________________ Total Score __________

SECTION 1 CONCEPT COMPREHENSION

Directions: Complete the following by writing the correct word or words in the respective blanks to the right.

With previous techniques, measurements or scores were used as data; however, chi square tests use ___(1)___ data. Further, because chi square can analyze nominal and ordinal variables, variance is not meaningful and thus the ___(2)___ are, for the first time, not calculated. When chi square uses one variable, the test is called a(n) ___(3)___-of-___(4)___ application. In such a case the ___(5)___ frequencies can generally be determined before data are collected. The number of individuals or frequencies that have been counted in each category are referred to as ___(6)___ frequencies and are the raw data used by chi square.

The calculated value of chi square is found by first subtracting the expected frequency of a category or cell from the observed frequency, ___(7)___ this difference, then dividing this result by the ___(8)___ frequency. These values obtained for all categories are then added together to get the ___(9)___ value of chi square. This procedure works except when there are too many expected frequencies of less than ___(10)___ or when df = ___(11)___, in which case ___(12)___ correction should be used.

When the chi square problem has two variables, the analysis involves a(n) ___(13)___ table. When the table's rows or columns are summed, the resulting totals are called ___(14)___, and these in turn are used in determining the ___(15)___ frequencies for the chi square. If the contingency table has R rows and C columns, the degrees of freedom = ___(16)___. In the case of a contingency table analysis, chi square is testing for a(n) ___(17)___ between the two variables involved.

As with other techniques that have been studied, the test of the null hypothesis involves comparing the calculated value of chi square with the ___(18)___ or tabled value of chi square. If the calculated value exceeds the tabled value, the null hypothesis is ___(19)___. In the case of both types of chi square, the general null hypothesis states that the expected frequency distribution does not differ from the ___(20)___ distribution.

Answers	For Scoring
1. ____________	1. ____
2. ____________	2. ____
3. ____________	3. ____
4. ____________	4. ____
5. ____________	5. ____
6. ____________	6. ____
7. ____________	7. ____
8. ____________	8. ____
9. ____________	9. ____
10. ____________	10. ____
11. ____________	11. ____
12. ____________	12. ____
13. ____________	13. ____
14. ____________	14. ____
15. ____________	15. ____
16. ____________	16. ____
17. ____________	17. ____
18. ____________	18. ____
19. ____________	19. ____
20. ____________	20. ____

SECTION 2 VOCABULARY PRACTICE

Directions: Identify the term defined and complete the designated squares.

Across

For Scoring

4. A two-dimensional chi square. ________
6. Chi square analyzes nominal data cell count or ______. ________
8. Data that are collected are referred to as the ______ frequencies in a chi square. ________

Down

1. The sum of the rows or columns of a contingency table. ________
2. E in chi square problems stands for ______ frequency. ________
3. The intersection of two sets or concurrent membership in two sets. ________
5. ______ statistical techniques do not require the computation of sum of squares. ________
7. If more than 20% of the expected values have frequencies of less than 5, ______ correction should be applied. ________
8. Yates' correction should be used for chi square problems with ______ df. ________

CHAPTER 16
Chi Square

Student ____________________ Score ________

SECTION 3 REHEARSAL EXERCISES

Directions: In the Answers column, write the letter that represents the correct choice.

	Answers	For Scoring
1. If a contingency table has R rows and C columns, the df for chi square are: (A) (R)(C) (B) (R − 1)(C − 1) (C) (R^2)(C^2) (D) R + C − 2	______	1. ______
2. For any particular alpha level, the critical value of χ^2: (A) increases as *N* increases (B) decreases as *N* increases (C) decreases with increases in df (D) increases with increases in df	______	2. ______
3. The area of rejection on a chi square sampling distribution lies: (A) in the tail to the right of the critical value (B) between the two critical values (C) in both tails beyond the critical values (D) near the middle of the distribution close to the mean	______	3. ______
4. In the chi square test of goodness of fit, if C equals the number of categories (levels) and *N* is the total frequency count, then the degrees of freedom are: (A) *N* − 1 (B) *N* − C (C) C (D) none of the above	______	4. ______
5. The chi square test of independence is used to test for: (A) differences between two or more means (B) differences between two sets of observed frequencies (C) similarity between expected values (D) a relationship between two variables	______	5. ______
6. In a chi square contingency test, the expected frequencies may be determined: (A) from the table of critical values of chi square (B) by multiplying the row marginal by the column marginal and dividing the product by *N* (C) by finding the difference between the observed and the critical values (D) prior to the collection of data	______	6. ______

(continued)

SECTION 3 REHEARSAL EXERCISES (continued)

Answers **For Scoring**

7. When the calculated value of the chi square statistic is less than the critical value: ______ 7. ______
   (A) conclude that the null hypothesis is tenable
   (B) conclude that what was observed is signficantly different from what would be expected
   (C) adjust chi square using Yates' correction
   (D) reject H_0

8. The chi square sampling distribution is: ______ 8. ______
   (A) discrete
   (B) normal
   (C) categorical
   (D) continuous

9. Applications of chi square goodness-of-fit tests and tests of independence involve how many variables? ______ 9. ______
   (A) one and two, respectively
   (B) two and two, respectively
   (C) two and one, respectively
   (D) one and one, respectively

10. How does the critical value of chi square with one df compare to the critical value of *F* with one and an infinite df for a particular alpha level, say 0.05? ______ 10. ______
    (A) Chi square is larger.
    (B) Critical *F* is larger.
    (C) They are the same values.
    (D) They are not comparable.

SECTION 4 PRACTICE PROBLEMS

Perform the following chi square analyses: **For Scoring**

1. Suppose 120 sophomores were asked their preference concerning six brands of soft drink. Preferences were as follows: 1. ______

	Brand					
	A	B	C	D	E	F
Observed	10	21	16	13	32	28

Test the null hypothesis that the soft drinks are equally preferred at the 0.01 level of significance.

____

____

____

____

____

(continued)

Student ______________________________ Score ____________

SECTION 4 PRACTICE PROBLEMS (continued)

For Scoring

2. In the field of genetics, Mendel's law states that in the pea family, round smooth is dominant over round wrinkled and yellow is dominant over green. According to the theory, the expected result of biological crossing is a 9:3:3:1 ratio of smooth and yellow, to smooth and green, to wrinkled and yellow, to wrinkled and green. In a crossing experiment the following results were tabulated:

2. ______

smooth and yellow	460
smooth and green	145
wrinkled and yellow	160
wrinkled and green	55

At the 0.05 level of statistical significance, do these data cast doubt on the theory?

3. Test for a relationship between rating on an important social issue and race using a chi square contingency test at the 0.05 level of significance. The data for the test are:

3. ______

	Race	
Rating	**White**	**Nonwhite**
Agree	35	18
Neutral	8	6
Disagree	20	12

(continued)

For Scoring

4. Students were classified by academic goals and asked whether they favored administration of minimum-competency exams for all students before graduation.

	Favor	No Opinion	Oppose
General Education	16	21	35
College Bound	28	10	5
Vocational	13	13	20

Use chi square to determine if the attitude toward such an exam for all students is independent of academic goal. Use the 0.05 level of significance.

4. ______

5. Rating of the school cafeteria by samples from the four grade levels resulted in the following tabulation:

	Freshman	Sophomore	Junior	Senior
Poor	6	8	10	4
Average	13	6	12	15
Good	18	9	12	7

Is the perception of the quality of the cafeteria food related to classification at the 0.05 level of significance?

5. ______

6. The following data represent scores on a 20-item personality assessment scale of a randomly selected sample of students from East Union School.

9	13	5	8	7	9	13	11	12	4	8	11	10
14	9	11	12	10	12	7	11	6	10	11	10	7
12	10	11	9	10	11	14	8	10	15	9	11	10
10	9	15	3	8	11	10	11	12	8	12	10	9
6	13	10	9	10	9	12	11	10	12	5	11	6
5	14	7	11	6	10	13	7	16	9	8	16	10
11	5	17	12	10	11	10	8	9	14	12	15	8
8	9	11	8	14	7	9	13	8	8	9	13	11
10	7	12	10	9	10	8	14	10	6	12	10	9
7	10	9	4	6	11	9	13	8	9	8	10	9

6. ______

(continued)

Student ____________________ Score ________

SECTION 4 PRACTICE PROBLEMS (continued)

The reference manual for the test states that the national norm for the test has a mean (μ) of 9.7 and a standard deviation (σ) of 2.8. Thus the national frequency curve has this appearance:

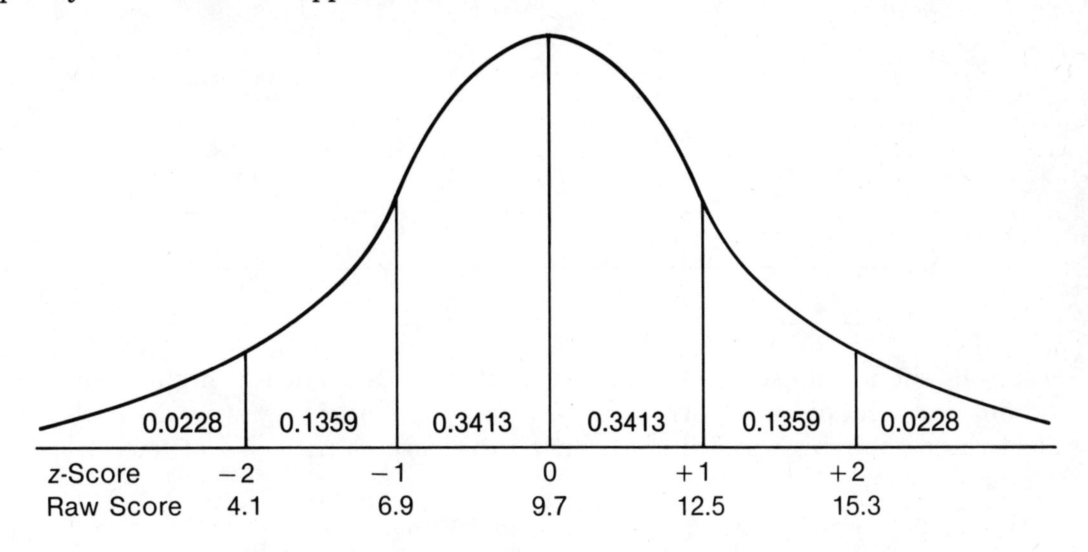

Determine the expected frequency of the six regions of the normal curve defined by the mean and ± 1 and ± 2 standard deviations from the mean. Then compute a chi square goodness-of-fit test and use the results to reach a conclusion about whether or not the East Union student body as a whole conforms to the national personality norm.

(continued)

For Scoring

7. Three editorials appeared in the school newspaper at Valley High School. A sample of readers were asked to indicate which editorial they considered to be the best. Of those sampled 215 preferred editorial A, 274 preferred editorial B, and 274 preferred editorial C. Is there sufficient evidence (at the 0.01 level of significance) to indicate a difference among the editorial evaluations?

7. ______

8. Data from a follow-up study of former students from a particular high school who did and who did not work at a part-time job while they were in high school rated their high school experience in retrospect with one of four descriptors as shown.

8. ______

Quality of High School Experience

	Excellent	Good	Fair	Poor
Did Not Work	28	35	59	36
Worked	42	56	43	26

Test the hypothesis that perceived quality of high school experience is related to whether or not the student worked while in high school. Use the 0.05 level of significance and interpret your results.

(continued)

Student ______________________________ Score __________

SECTION 4 PRACTICE PROBLEMS (continued)

For Scoring

9. A sample of students were asked to indicate their grade level in school and to check which volunteer service activity they had participated in during the past year. 9. _____

Grade Level	Blood Donor	Community Clean-up
Freshman	12	14
Sophomore	16	15
Junior	19	12
Senior	26	10

Determine if grade level is related to the type of volunteer service rendered. Interpret these results.

____

____

____

____

____

____

____

10. Is gender related to preference for ice cream flavor according to the following data? 10. _____

	Student Gender	
Brand Preferred	Male	Female
Vanilla	16	14
Chocolate	24	20
Strawberry	17	16

____

____

____

____

____

____

____

(continued)

For Scoring

11. The two statistics classes at Maryville High are open to students in any grade level, freshman through senior. Test to see if this year's enrollment pattern is similar to the previous year's enrollment relative to the distribution across grade levels.

11. _____

	Freshman	Sophomore	Junior	Senior	Total
This Year	5	7	13	22	47
Last Year	15%	20%	28%	37%	100%

[Hint: To compute the expected frequencies (last year's distribution), one must adjust the percentages by multiplying each of last year's percentages by 47; thus the expected frequency for freshmen would be $0.15 \times 47 = 7.05$. Continue this strategy for the other three grade levels.]

12. A sample of 120 students were classified by whether or not they belonged to a school-sponsored social club and by their grade level. Do the data suggest that membership in a social club is independent of grade level?

12. _____

	Freshman	Sophomore	Junior	Senior
Belong	12	14	18	19
Do Not Belong	18	14	12	13

Interpret your findings.